**Illustrators**
Mike Atkinson, Jim Dugdale, Ron Jobson,
Janos Marffy, John Marshall, Mike Saunders

**Author**
John Paton

# The CHILDREN'S
# FIRST SCIENCE
# ENCYCLOPEDIA

Compiled by Michael Dempsey

Exeter Books

NEW YORK

# Contents

# Stars and Galaxies

If you look up at the sky on a clear night you will see thousands of stars – each a small twinkling point of light. Yet many of these stars are ten thousand times more powerful than our Sun. Most of the stars we see belong to our galaxy – the Milky Way. The Milky Way is a huge flat disc of stars – million upon million of them – all going round and round like a giant bicycle wheel. Our Sun is a quite ordinary star lying about two-thirds of the way out from the center. The Sun travels around the hub of the galaxy at a speed of 155 miles per second. But the Milky Way is so vast that it takes the Sun 225 million years to make just one trip all the way round.

Other galaxies of stars exist far out beyond our own Milky Way. As far into space as astronomers can look with their biggest telescopes, more and more galaxies come into view. And each of these galaxies contains thousands of millions of stars. These great galaxies are so far away they appear

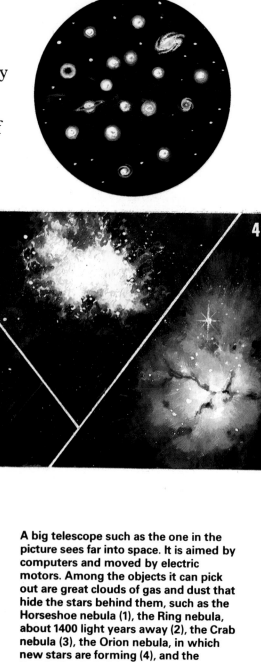

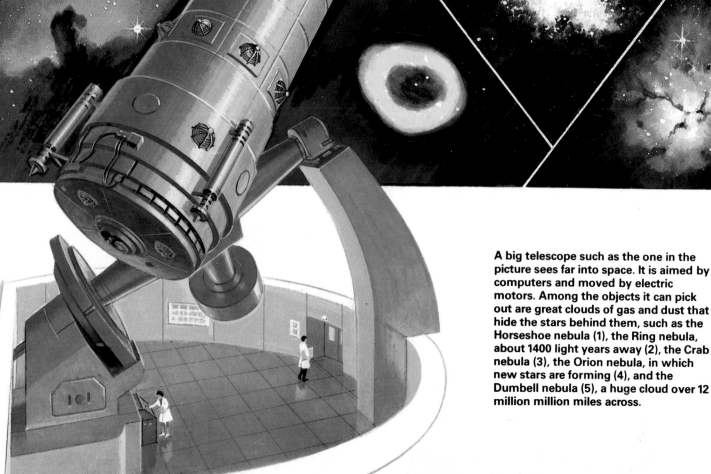

A big telescope such as the one in the picture sees far into space. It is aimed by computers and moved by electric motors. Among the objects it can pick out are great clouds of gas and dust that hide the stars behind them, such as the Horseshoe nebula (1), the Ring nebula, about 1400 light years away (2), the Crab nebula (3), the Orion nebula, in which new stars are forming (4), and the Dumbell nebula (5), a huge cloud over 12 million million miles across.

A star such as the Sun is born when a cloud of gas and dust contracts and starts to glow. There are several stages in the life of a star. Gravity first of all pulls the cloud material together and makes it heat up. This may take 20 million years. Nuclear fusion reactions begin inside the star and keep its temperature steady for thousands of millions of years. Then many stars begin to swell up, the surface cools and they become *red giants* or supergiants. Our Sun will not reach its red giant stage for thousands of millions of years. After millions more years it will shrink once more to become a small very dense *dwarf star* about the size of the Earth. It will be so dense that a tablespoonful of its matter will weigh several tons.

dimmer than single stars in our own galaxy. The light we see from some of them has taken 8,000 million years to reach us, and light travels at a speed of 186,000 miles every second!

And the astronomers have discovered a strange thing. The distant galaxies are all flying away from us and from each other. The further away they are, the faster they appear to be flying apart. Back-tracking the flight of the galaxies, astronomers have worked out that they must all have been close together or all joined up about 20,000 million years ago. This may have been the time of the Big Bang when all the universe began.

Astronomers are puzzled by strange objects called *quasars*. These bodies on the edge of space throw out very powerful radio and light waves – so powerful that their light is greater than a hundred galaxies, even though their size is quite small. Perhaps the quasars will help us to find out how our universe came into being.

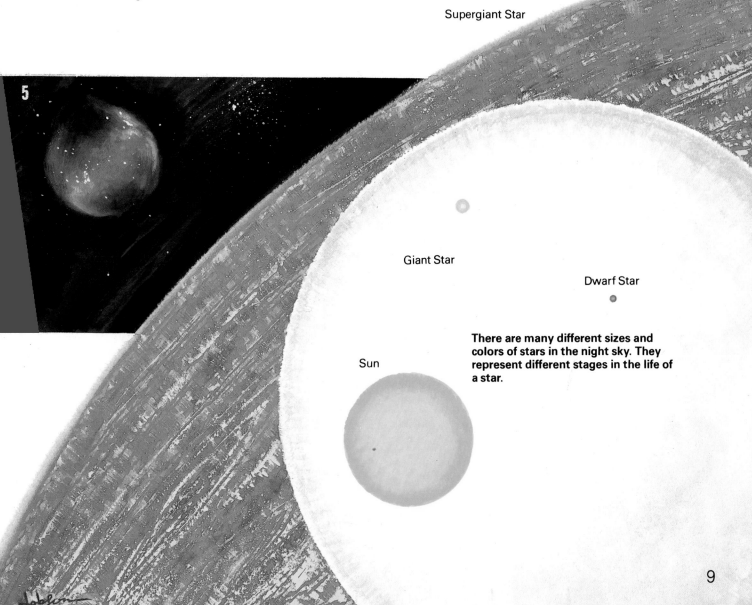

Supergiant Star

Giant Star

Dwarf Star

Sun

There are many different sizes and colors of stars in the night sky. They represent different stages in the life of a star.

1    2    3    4    5

Sun

# *Family of the Sun*

Our planet Earth speeds through space on its yearly journey around the Sun. Eight other planets, their moons and smaller lumps of rock and ice are also traveling endlessly around the big, hot Sun. All these put together we call the Solar System – 'solar' means 'of the Sun'.

The Sun is a huge ball of very hot gas, so big that more than a million Earths could be fitted inside it. It is the Sun's *gravitation* – its pull through space – that keeps all the planets and other bodies in the Solar System traveling around it.

Our home planet, Earth, takes just over 365 days to go once round the Sun. We call this a year. The Earth is also spinning like a top. It takes 24 hours to go round once. This is a day. It is the Earth's daily spin that makes the Sun and the stars appear to rise and set. All the other planets go round the Sun and spin, just as Earth does. But they all spin at different speeds and orbit the Sun in different lengths of time. The far-away planet Pluto takes about 250 years to go once around the Sun. Little Mercury, the closest planet to the Sun, goes around it in only 88 days.

The Sun is a huge ball of very hot gas, about 857,000 miles across. But it is only one of millions of stars in the universe. It looks big to us because it is so close. The next nearest star to us is about 250,000 times as far away as the Sun. Dark spots can often be seen on the Sun's surface. These sunspots come and go and their number varies from year to year. They look dark because they are cooler than the surrounding hot gas. The Sun also shoots out huge flames from its surface. These *prominences* can stretch out for a million miles into space before falling back into the Sun.

Never look at the Sun, either with the naked eye or through the lens of a telescope or binoculars.

6                                        7            8            9

The more scientists find out about the other planets in the solar system, the more we realize what a wonderful place Earth is (3). Everything about Earth is just right for us. It is the right distance from the Sun. If it were as close as Venus, the heat and the choking carbon dioxide gas would have built up a thick, boiling hot atmosphere. Mars is our closest neighbor, but if the Earth was as distant from the Sun and as small as Mars water vapor would turn to ice and most of the atmosphere would vanish into space. If Earth were as far away from the Sun as Neptune, the whole planet would be nothing but frozen hydrogen surrounding a core of rock and ice. The picture above shows what it might be like on Neptune.

### PLANET FACTS

*Mercury*, the smallest planet, is not much larger than our Moon. It is over 3,000 miles across, the Moon is 2,160 miles (1).

*Venus,* the second planet, is the brightest object in the sky after the Sun and Moon. Scientists think Venus is so bright because it is surrounded by an unbroken layer of white clouds (2).

*Mars* is called the Red Planet because its surface rocks are reddish. Viking space probes landed on Mars but could find no sign of life on the planet (4).

*Jupiter* (5) is the biggest planet. Its surface is made up of swirling clouds with one big red spot always there.

*Saturn* is well known for the rings that encircle it. The rings are made of tiny ice fragments (6).

*Uranus* (7) and *Neptune* (8) are distant icy planets. They look like small greenish discs when seen through even the largest telescope.

*Pluto* (9) is usually the planet furthest from the Sun. Some astronomers think it may be even smaller than Mercury. Its orbit is so strange that at the moment it is closer to the Sun than Neptune.

# The Earth in Space

The Earth spins around on its axis once every 24 hours. Its axis is an imaginary line that goes through the North and South Poles. But because the Earth's axis is tilted, we have seasons. If the axis were straight up and down in relation to the Sun, we would get the same amount of sunlight every day. There would be no spring, summer, fall and winter.

Towards the end of June, the northern part of the Earth tilts most towards the Sun. Around the North Pole it is sunlight for 24 hours a day. The southern parts of the Earth get less sunlight at this time because the Sun's rays strike the southern lands at a slant. At the South Pole it is dark all the time. Six months later, the Earth's axis tilts the other way. At the end of December, the seasons are reversed. Southern lands get most sunshine, northern lands get very little. It is summer in Australia and winter in Britain.

Twice a year, at the end of March and September, the Sun shines directly over the equator. All parts of the world have equal night and day.

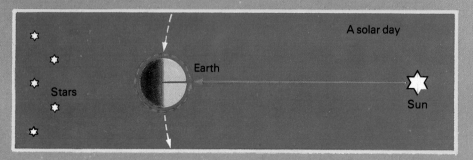

A solar day

Stars

Earth

Sun

The movements of the Sun, planets, their moons and the stars are very complicated. The Moon moves round the Earth; the Earth moves round the Sun, the Sun moves round our galaxy. And they all spin on their own axes. We set our clocks according to the time it takes the Earth to spin around once on its axis *relative to the Sun* – 24 hours. But during those 24 hours the Earth also moves on its journey round the Sun. And it also moves relative to the stars.

The Earth revolves once, relative to the stars, in 23 hours 56 minutes – four minutes less than the solar day. This means that the stars appear to rise and set four minutes earlier each day. This adds up to 24 hours a year. The stars rise and set at a given hour only once a year, on or near the same date. This is when the Earth has returned to the same point in its trip around the Sun. To make their jobs easier, astronomers often use something called *siderial time* – star time. In this siderial time scale, the stars rise and set at the same time every day.

### How the Planets Began

Astronomers have wondered for many years how the planets came into being. Most of them now believe that the planets formed from the vast cloud of gas and dust that produced the central Sun. The cloud consisted mostly of hydrogen, the commonest element in the universe. Over millions of years the gas condensed in clumps. Heavy elements like iron and nickel formed at the core of each planet. Above this central core floated lighter elements that became the rocks of the Earth's crust. Thus our Earth was born about 4,600 million years ago. First it was an inferno of molten red-hot rock surrounded by burning gases. Gradually it cooled down, rain filled the oceans and Earth became the planet that we know.

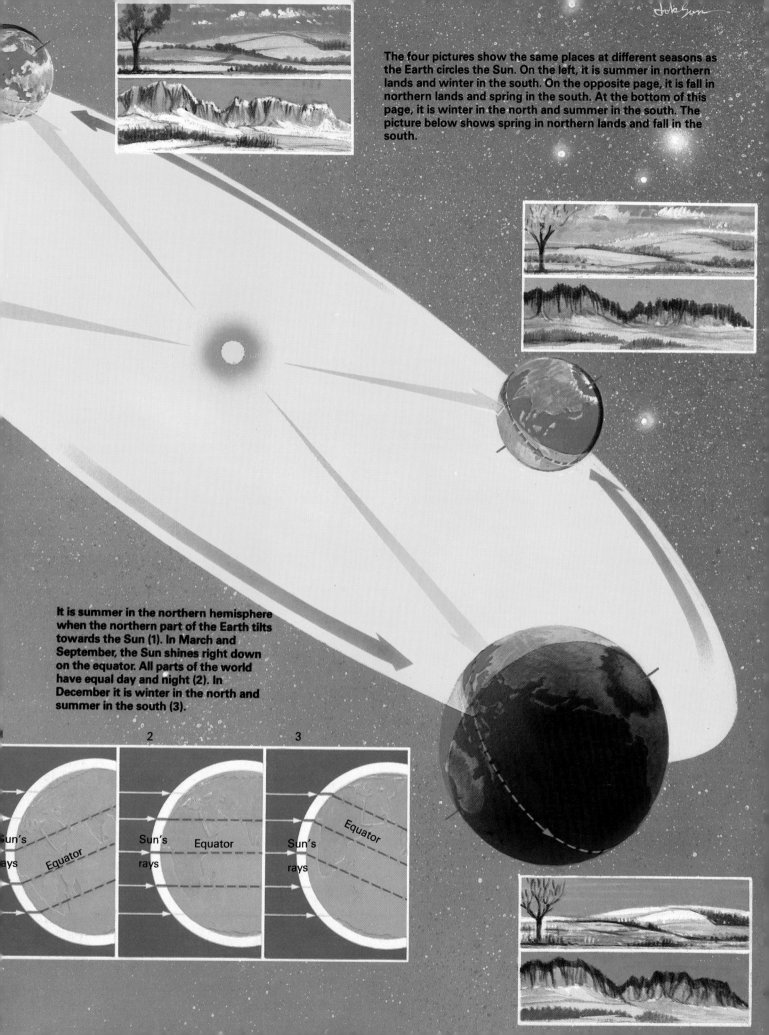

The four pictures show the same places at different seasons as the Earth circles the Sun. On the left, it is summer in northern lands and winter in the south. On the opposite page, it is fall in northern lands and spring in the south. At the bottom of this page, it is winter in the north and summer in the south. The picture below shows spring in northern lands and fall in the south.

It is summer in the northern hemisphere when the northern part of the Earth tilts towards the Sun (1). In March and September, the Sun shines right down on the equator. All parts of the world have equal day and night (2). In December it is winter in the north and summer in the south (3).

2

3

Sun's rays

Equator

Sun's rays

Equator

Sun's rays

Equator

# The Moon

The Moon is our nearest neighbor in space. It travels around the Earth at a speed of about 2,280 miles per hour. It orbits the Earth once every month or so. The Moon, like the Earth, is constantly spinning. It takes it a month to complete one spin. This means that we always see the same face of the Moon from Earth.

When the night sky is clear, we are often surprised how brightly the Moon shines. But the Moon doesn't shine with its own light. It is merely reflecting light from the Sun. Astronauts on the Moon see Earth as a big ball of light. Again, the light from Earth is only reflected sunlight.

Seen from Earth, the Moon looks about the same size as the Sun. But this is only because the Moon is so close to us. In fact, the Moon is quite small – about the same size as Australia. It is so small that it does not have enough gravity to hold an atmosphere around it. There is no air on the Moon, so spacemen have to take their own air with them. And the Moon has no weather – no wind. During the day, the Sun beats down with an intense heat of about 250°F – hotter than boiling water. During the lunar night, the temperature falls to a freezing –250°F.

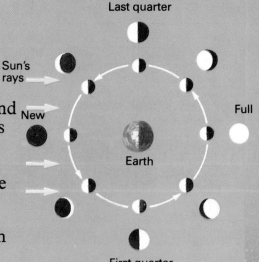

Last quarter

Sun's rays

New

Earth

Full

First quarter

The Moon goes through a 29½-day cycle of 'phases' as it circles the Earth. These phases happen because we see only the half of the Moon that is illuminated by the Sun. At New Moon, the Moon cannot be seen because its dark face is turned towards the Earth. After two or three days, it has moved far enough to be seen as a thin crescent in the sky. After seven days it is a perfect half circle called the First Quarter. A week later it is full. We see the full sunlit circle as it lies opposite the Sun in the sky. After this, the phases go into reverse – the Moon wanes until it disappears once more.

Sun's rays

Moon

Shadow

Earth

It is strange that the Sun and Moon appear in the sky to be the same size although the Moon is really 400 times smaller. It is just much closer to us. The Moon sometimes passes in front of the Sun, blotting out the Sun's light. This is a *solar eclipse.* The Moon's shadow is a cone, as you can see in the picture. To see a total eclipse we must be inside that cone. When the tip of the shadow cone reaches Earth it is only about 150 miles across. It is only in this 150 mile-wide circle that people see a total eclipse of the Sun.

The Moon has many of the valuable minerals that are becoming scarcer and scarcer on Earth. Perhaps in years to come the Moon's minerals will be mined and transported as in the picture below. Lunar gravity is only one-sixth of Earth's, so it is much easier to shoot things off its surface into space. The minerals could be put in huge buckets and fired off a track by magnetic waves. This mechanism is called a mass driver. Out in space, the minerals could be 'caught' by a space tug (right) and taken to wherever they are needed.

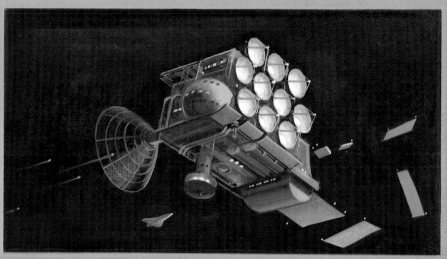

15

# *Journeys into Space*

On March 16, 1926, an American scientist, Dr Robert Goddard, fired the first liquid-fueled rocket into space. Goddard's rocket rose only 200 feet into the air, but this was the first tiny step towards bigger and more powerful rockets that took people to the Moon and probes right out of our Solar System.

A mere 31 years after Goddard's rocket, the Russians launched Sputnik 1, the first man-made satellite to orbit the Earth. This happened on October 4, 1957, and it was on that day that the Space Age began. Since then, hundreds of satellites have been fired into space and the Shuttle blasts off into space and lands back on Earth with an ease we are now beginning to take for granted.

## Probes to the Planets

Some of the most exciting space projects have been the unmanned probes sent up to find out more about other planets in our Solar System. In 1976, two *Viking* spacecraft landed on Mars. They carried out tests to find out whether any form of life existed on the Red Planet; but they could find none.

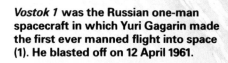

*Vostok 1* was the Russian one-man spacecraft in which Yuri Gagarin made the first ever manned flight into space (1). He blasted off on 12 April 1961.

The Americans who landed on the Moon traveled in *Apollo* craft (2). The Command Module for the men was in the nose of the spacecraft.

*Venera 4* (3) was one of several Russian craft that reached Venus and released a capsule by parachute (right). The *Veneras* radioed back to Earth several important discoveries.

American *Pioneer* probes (4) flew close to Jupiter in 1973 and 1974. They took some remarkable close-up pictures of the giant planet.

A *Viking* Mars craft is seen at (5).

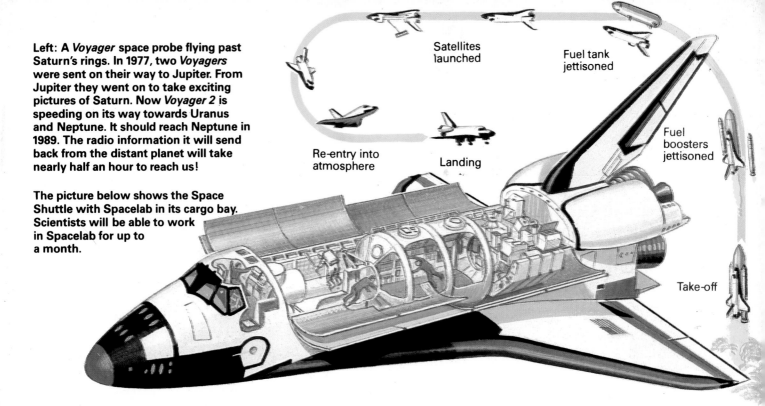

Left: A *Voyager* space probe flying past Saturn's rings. In 1977, two *Voyagers* were sent on their way to Jupiter. From Jupiter they went on to take exciting pictures of Saturn. Now *Voyager 2* is speeding on its way towards Uranus and Neptune. It should reach Neptune in 1989. The radio information it will send back from the distant planet will take nearly half an hour to reach us!

The picture below shows the Space Shuttle with Spacelab in its cargo bay. Scientists will be able to work in Spacelab for up to a month.

Satellites launched

Fuel tank jettisoned

Fuel boosters jettisoned

Re-entry into atmosphere

Landing

Take-off

3

4

5

## The Space Shuttle

When the Space Shuttle *Columbia* blasted off in 1981 a new kind of space travel began. After a flight of 54½ hours in space, it glided back to a perfect landing on a desert airstrip.

Before the Shuttle, all rockets and manned spacecraft were used only once. This made space flight very expensive. The Space Shuttle is a combined launch rocket and spacecraft that can be used many times.

The Shuttle is launched by rocket like other spacecraft, but it glides back to land on a runway like an aircraft. It has three main rocket engines which are fed with fuel from a big tank. This tank is dumped when all its fuel is gone. Two extra rockets are attached to the sides of the Shuttle to help it into space. These rockets fall away as the ship climbs, and drop by parachute into the ocean. There they are recovered and can be used again. As the Shuttle glides back through the atmosphere, special tiles protect it from the fierce heat.

17

# Cities in Space

The biggest problem to face the world within the next few decades may be how to feed and house all its people. The world's population is growing fast, especially in Third World countries. Before the year 2080 the population of our Earth may be three times its present size. Unless we can halt the population explosion and find new sources of food and raw materials, people may have to move out into space.

There have been many designs for giant space colonies with room for thousands of people. These colonies would be vast cylinders or spheres, going round and round to give Earth-like gravity for the people who live there.

## Island Three

One idea for a space colony is called *Island Three*. It would be made up of two or more huge cylinders positioned at a fixed distance from the Earth and Moon. The cylinders are nearly 19 miles long and about 4 miles around. Giant mirrors reflect sunlight through three long windows in the sides of the cylinders. The angle of the mirrors can be altered to make artificial day and night and temperature changes. Even changing seasons can be created. Inside the cylinders will be three long land areas where people live and grow crops, just as they would on Earth.

*Island Three* **will have a solar power station at one end and a ring of factory units at the other. There will also be vast docking facilities for spaceships.**

## Artificial Gravity

To make life possible for people who go to live in space, there must be gravity. Gravity is the force that pulls us all towards the centre of the Earth – that gives us weight. Out in space there is no gravity. People and things just float around.

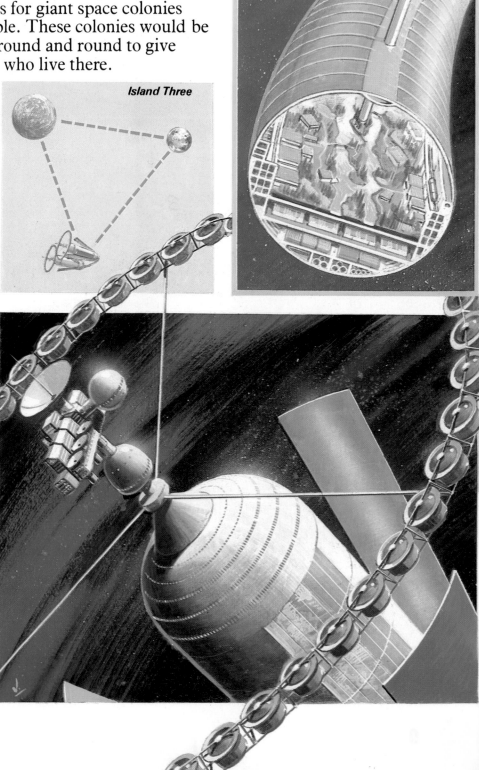

*Island Three*

Artificial gravity can be made by spinning a whole space colony at just the right speed. The force of the spin pushes everything towards the outside. To anyone in a space colony, 'up' will be towards the inside. Two people on opposite sides of the craft will be upside down in relation to each other.

## A Torus Colony

Another idea for a space colony is the building of a huge torus or wheel, like the one shown here. The wheel is rotated around the central hub once every minute. This creates an artificial gravity that holds all the land, water and people in place inside the vast tube. The tube itself is over 6,500 feet across. On the central hub is a solar power station that provides energy to run the whole colony. Also in the hub are factories where many of the colonists work, perhaps using materials mined on the Moon. Life in the tube would be made as much like life on Earth as possible for the colonists.

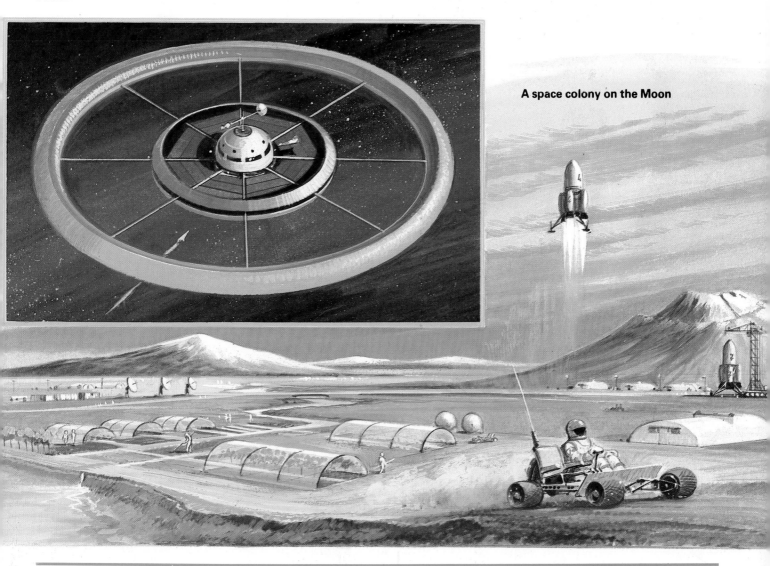

A space colony on the Moon

## Will we go to other Stars?

In the future, people may be able to go to other star systems outside our Solar System. But at present we do not know how it will be possible. The nearest star to us, other than our Sun, is Alpha Centauri, and it would take a present-day rocket ship about 100,000 years to get there! So we will have to develop much faster rocket engines than those we have at present.

But we will almost certainly begin by putting people on other planets and moons in our Solar System. The picture above shows a colony on the Moon. Life will not be too easy for the first settlers. It will take them some time to get used to the very small gravity and the need to carry their own air around with them. Water will either have to be imported from Earth or manufactured on the Moon. Food will be grown in large greenhouses.

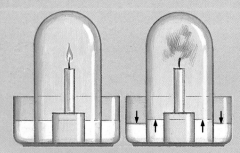

# The Air We Breathe

Nothing can burn without oxygen. We can prove this by a simple experiment. Light a candle and stand it in a bowl of water. Place a jar over it. As the candle burns, the oxygen in the jar is used up and the water slowly rises inside the jar to take its place. Then the candle goes out. The oxygen has been used up.

Our planet is surrounded by a blanket of air we call the *atmosphere*. It stretches upwards for several hundred miles and it contains gases that all living things must have. We are all in contact with air every second that we live, but we are seldom aware of it. It is quite invisible and has no taste or smell. You can feel the air when the wind blows. You can see clouds being pushed along by the air. Air can turn windmills. And if there were no air we would live in a silent world. Sound needs air to travel through. It cannot travel in a vacuum.

Although the atmosphere is hundreds of miles thick, over three-quarters of all our air is in the few miles nearest to Earth. As we go higher, the air grows thinner and thinner. At the top of a high mountain, there is so little air that we have difficulty breathing. There is not enough life-giving oxygen in each breath we take. That is why people who climb Mount Everest take their own oxygen supply with them. Inside airliners, the air pressure has to be kept as it is on Earth so that people can breathe normally.

Although we cannot see air, it is a substance just as rocks are. It is pulled down towards the Earth by the force of gravity – it has weight. When we talk about air pressure, we are talking about the weight of air pressing down on us, and air has quite a lot of weight. It presses on every square inch of our bodies with a force of over 70 ounces.

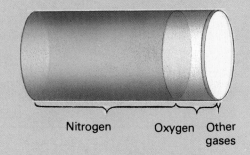

Nitrogen          Oxygen   Other gases

In the experiment with the candle and the jar, it was found that extra water sucked into the jar was about one-fifth of the jar's volume. This shows that about one-fifth of the air is oxygen. Air is made up of a mixture of gases that we cannot see. Over three-quarters of it is nitrogen (78 per cent). As oxygen makes up about one-fifth (21 per cent), this leaves only 1 per cent, which is made up of small quantities of other gases such as *argon, helium, carbon dioxide, hydrogen, ozone,* etc. Air also contains some water vapor – separated particles of water too fine to see. When we talk about *humidity* we are talking about the amount of water vapor in the air. When the air holds as much water vapor as it can hold without mist appearing, we say the humidity is 100 per cent.

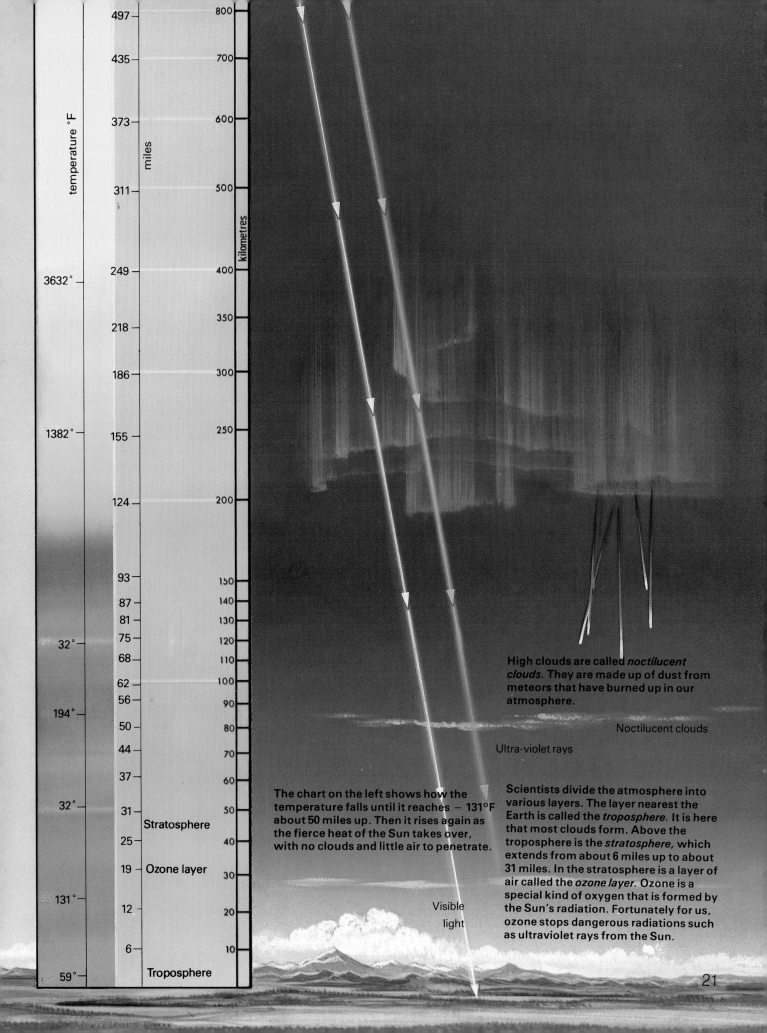

temperature °F

497
435
373
311
3632°
249
218
186
1382°
155
124
93
87
81
32°    75
68
62
194°   56
50
44
37
32°    31
25
19
131°   12

6

59°

miles

kilometres

800
700
600
500
400
350
300
250
200

150
140
130
120
110
100
90
80
70
60
50
40
30
20

10

Stratosphere

Ozone layer

Troposphere

High clouds are called *noctilucent clouds*. They are made up of dust from meteors that have burned up in our atmosphere.

Noctilucent clouds

Ultra-violet rays

The chart on the left shows how the temperature falls until it reaches − 131°F about 50 miles up. Then it rises again as the fierce heat of the Sun takes over, with no clouds and little air to penetrate.

Visible light

Scientists divide the atmosphere into various layers. The layer nearest the Earth is called the *troposphere*. It is here that most clouds form. Above the troposphere is the *stratosphere,* which extends from about 6 miles up to about 31 miles. In the stratosphere is a layer of air called the *ozone layer*. Ozone is a special kind of oxygen that is formed by the Sun's radiation. Fortunately for us, ozone stops dangerous radiations such as ultraviolet rays from the Sun.

21

# Weather in the Making

Life is only possible because of the layer of air wrapped around our Earth – the layer we call the atmosphere. The atmosphere protects us from the Sun's fierce rays and gives us air to breathe. It also gives us our ever-changing weather.

Weather depends on the movement of air we call winds. Movement of air is caused by differences in the temperature of the air. When air is heated or cooled, it moves.

Several things cause temperature differences in the air. Some parts of the Earth get more heat from the Sun than others. Because the Earth is curved, the Sun's rays are

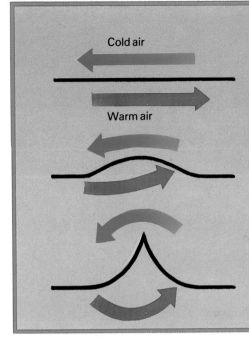

Cold air

Warm air

The diagram on the right shows why the Sun's rays do not heat the Earth evenly all over. Because the Earth is a ball, the rays come straight down on the equator. At the poles they arrive at a slant, so the heat is spread over a larger area.

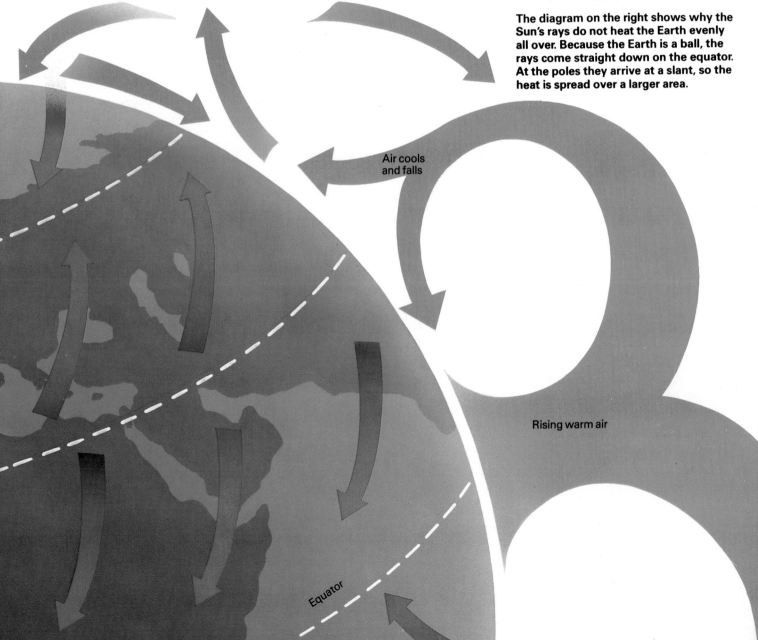

Air cools and falls

Rising warm air

Equator

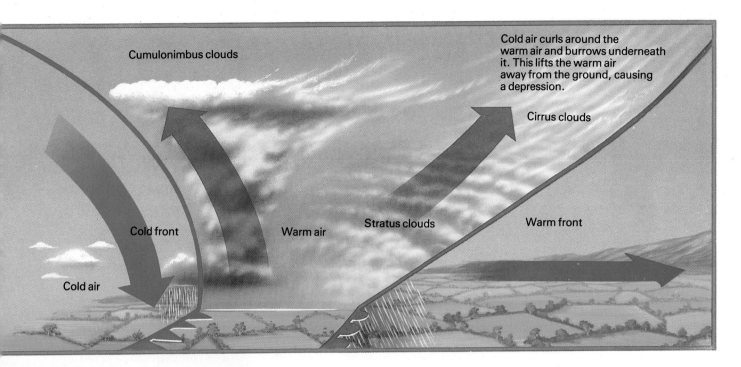

Cumulonimbus clouds

Cold air curls around the
warm air and burrows underneath
it. This lifts the warm air
away from the ground, causing
a depression.

Cirrus clouds

Cold front

Warm air

Stratus clouds

Warm front

Cold air

The diagram on the left shows some of
the main air movements round the
Earth that give us our weather. Rising
warm air at the equator moves off
towards the North and South Poles. As
the air cools in the colder atmosphere
high up, it falls again and is sucked back
towards the equator. There it takes the
place of the rising warm air. Other air
from the equator goes on towards the
poles and forms other circular
movements of air masses. In addition,
the rotation of the Earth drags the winds
to the west. Other air movements are
caused by differences between land and
water and how well they hold heat. The
changing seasons also have an
important effect on the great air masses.

**Periods of unsettled weather, with rain, gales and sometimes snowstorms are
caused by centres of low air pressure called *depressions*. Depressions happen when
cold air from the poles meets warm air from the tropics. The cold air curls around
the back of the warm air and a depression is made.**

strongest at the equator. At the cold poles, the same amount
of Sun heat is spread over a bigger area. So the air around
the equator is warmer than the air around the poles. Warm
air rises. The rising air at the equator moves off towards the
North Pole and the South Pole. It also becomes cooler as it
goes up, so part of the way to the poles some of the air sinks
and returns to the equator. This makes a circular movement
of air.

There are other air movements between the poles and the
equator. In addition, the Earth's rotation pushes the winds
to the west. But this wind pattern is made more complicated
because the Earth's surface is made up of land and sea.
Land heats up more quickly than water; it also loses its heat
more quickly. This causes differences in the air temperature
over various parts of the world and therefore air movements.
The changing seasons further complicate the weather
pattern.

All these things cause great masses of air to wander about
the Earth's surface. It is these wandering air masses that give
us our changing weather. They meet up with each other,
they rise and fall and, of course, they carry rain. People who
study weather and make forecasts take measurements of
temperature, pressure, wind force and direction and
humidity (the amount of water in the air). They draw
weather charts and make predictions with the help of big
computers.

# Continents Adrift

Only about a quarter of the surface of our planet is dry land. The rest is sea. Most of the Earth's land lies north of the equator and is broken up into the masses we call 'continents'. But it was not always so. Over 200 million years ago, when the first dinosaurs were beginning to roam the world, all the Earth's land was joined together in one huge mass. This great land mass has been called Pangaea. Over millions of years, Pangaea moved and broke up to form the continents as we know them. This movement is still going on at a rate that varies between ½ inch and 5 inches a year. It is called the 'continental drift'.

If we compare the shapes of the coasts of western Africa and eastern South America we can see that they fit together quite well. And if these continents are matched, not on the shore lines but at their under-sea *continental shelves*, the jigsaw fit is better still. (All the continents have under-sea shelves sloping out from them.) America and Africa were once joined together.

The picture on the left shows what happens inside a volcano. The deeper you go under the Earth's surface, the hotter it gets. At a depth of about 20 miles it begins to get so hot that some rocks simply melt. This molten rock is called *magma*. Some of this magma is pushed up through cracks and holes. These are volcanoes. There are different kinds of volcanoes. Some erupt quietly, oozing out molten rock called *lava*. The lava may spread out for miles before its cools and hardens. Lava of this kind builds gradually sloping mountains. Explosive-volcanoes throw out rocks mixed with gas and steam that has been trapped underground.

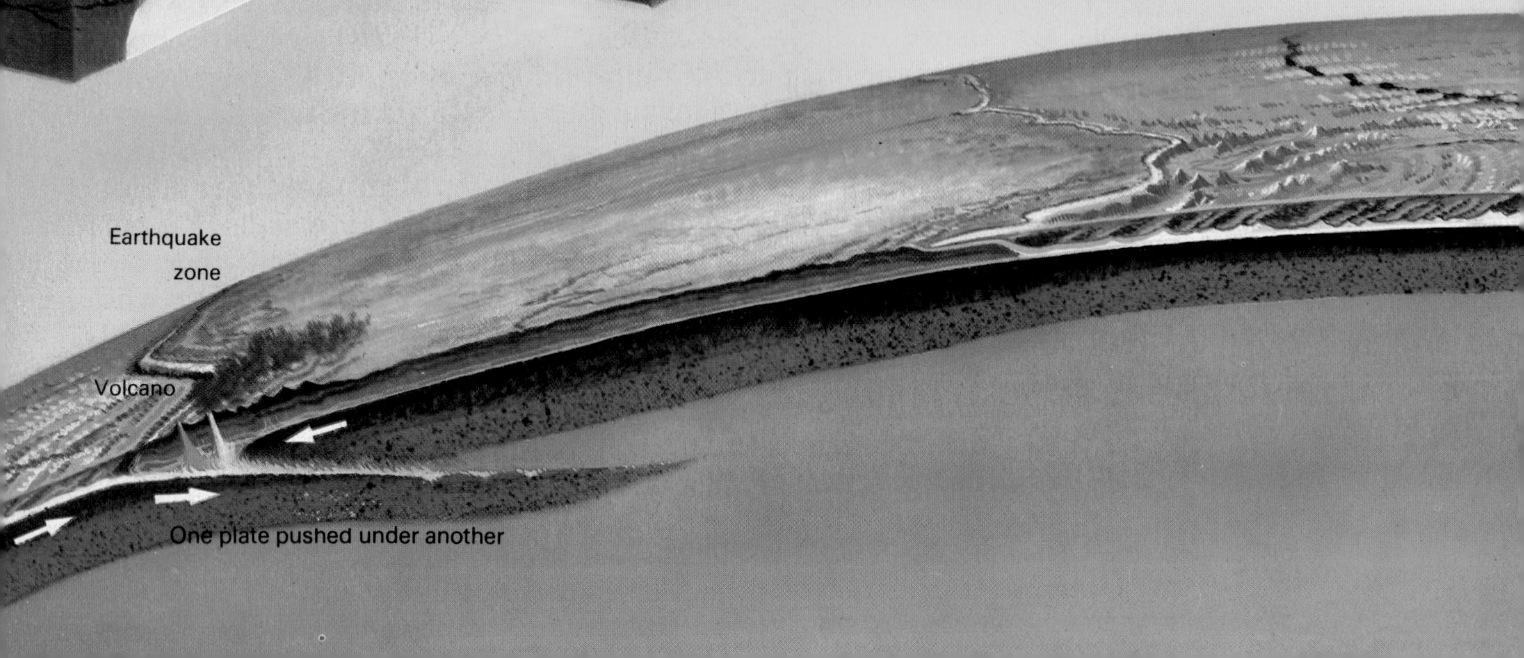

Lava

Magma

Earthquake zone

Volcano

One plate pushed under another

The Earth's crust beneath our feet is made up of two main kinds of rock. Great blocks of granite-type rock, which we call the continents, are embedded in a heavier kind of hot, half-liquid rock. The continents are great plates 'floating' on the hot rock underneath. They move very slowly, but the huge mass of the plates means that they move with tremendous force. When two plates come towards each other, the great force of the meeting pushes one plate under the other. In plate collisions such as this, mountain ranges are slowly pushed up and there may be earthquakes.

In the course of time, the continents have traveled enormous distances. By examining fossils in the rocks, and by other means, scientists are able to plot the history of a place's climate. They know, for instance, that frozen Antarctica was at one time in the tropics. And it is possible to tell the likely future movements of the continents. Africa, for example, will drift north, slowly closing the Mediterranean Sea. Australia will continue its slow journey northward. And by measuring magnetic field directions fixed in rocks of different ages, experts have been able to plot the drift of Britain's North Sea oil rocks from the time when they were south of the equator 400 million years ago.

The theory of the drifting continents has also helped to explain how closely related animals are found in lands now separated by thousands of miles of sea.

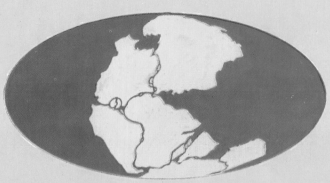

200 million years ago

60 million years ago

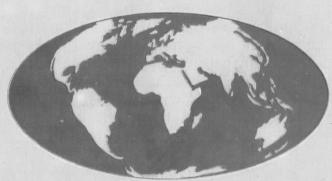

Today

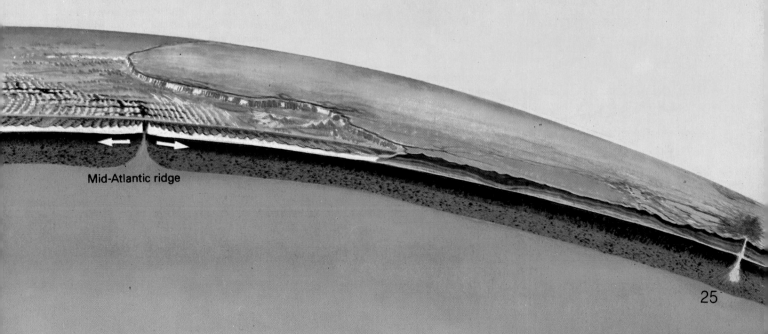

Mid-Atlantic ridge

# Rocks and Minerals

Much of the Earth is rock. Mountains and hills are made of rock. The soil is mostly fine rock particles. Stones and pebbles are small pieces of rock. And all the rocks in the Earth's crust are made up of substances called minerals. There are thousands of different minerals and, like all substances, minerals are built up from chemical elements. There are only about a hundred elements – substances such as oxygen, iron and carbon. Most minerals are made up of mixtures of several elements. But a few have formed from only one element. Diamond, for example, is a pure form of carbon.

The most common elements in the Earth's crust are oxygen and silicon. Quartz, the most common mineral, is a mixture of these two elements.

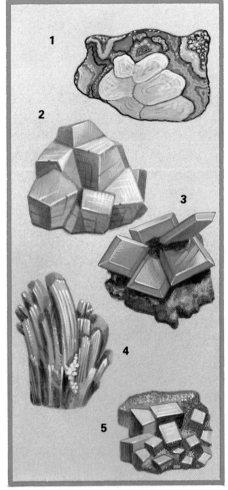

## The Three Types of Rock

There are three main types of rock in the Earth's crust. They are called *igneous*, *sedimentary* and *metamorphic* rocks. Igneous rocks formed from hot molten material from inside the Earth. Sedimentary rocks are made up of tiny particles including the remains of fossils of microscopic creatures that lived in the sea millions of years ago. As these creatures died, their shells dropped to the sea floor and piled up and were squeezed until they became rock. Limestone, sandstone and chalk are sedimentary rocks. Metamorphic rock has been made by the changing of existing rock by heat or pressure. Marble is a metamorphic rock.

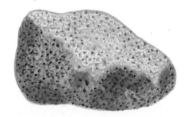

Sandstone is a common sedimentary rock. The wearing away of sandstone makes up a large part of our beaches. The main ingredient in sand is quartz.

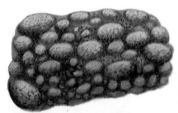

Conglomerate is a mixture of rock fragments cemented together by finer particles. The pebbles in conglomerate are any hard rock such as flint or quartz.

Chalk is a soft, white limestone. It was formed as mud on the bottom of an ancient sea. Chalk consists mainly of tiny shells and calcite crystals.

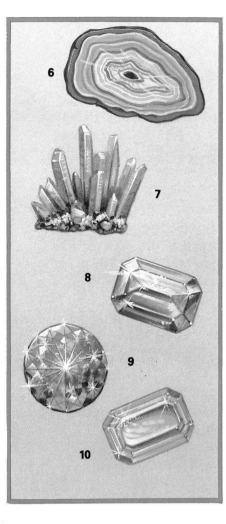

6

7

8

9

10

## Minerals

The panels on the left show some of the many minerals that are found in the crust of the Earth. The minerals in the far left panel contain metals. Minerals that contain sufficient metal for it to be extracted are called ores. The minerals in the near left panel include semi-precious stones and gems. Most of them do not contain metals.

The minerals pictured are: (1) Malachite, an important and beautiful ore of copper; (2) Galena, the chief ore of lead; (3) Wulfenite, a strikingly colored ore of the valuable metal molybdenum; (4) Zincite, an ore of zinc; (5) Iron pyrites, also known as fools' gold because its color has led prospectors in the past to believe they have struck gold; (6) Agate, an attractive form of quartz; (7) Quartz crystals; (8) Amethyst, another form of quartz, widely used in jewelry; (9) Diamond, an extremely pure form of carbon; (10) Emerald, an extremely complex mineral containing beryl and aluminum. The last three minerals are shown cut and polished as they are normally seen in jewelry.

Fossil prints left in ancient rocks tell the scientists much about the past. This is a fossil of a fern plant that lived about 250 million years ago.

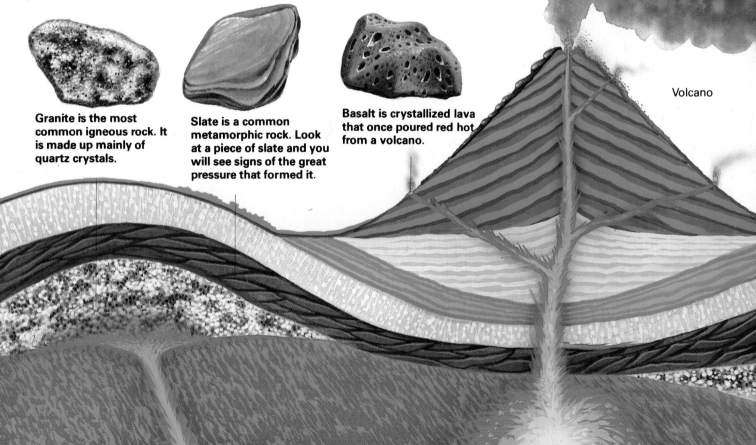

Granite is the most common igneous rock. It is made up mainly of quartz crystals.

Slate is a common metamorphic rock. Look at a piece of slate and you will see signs of the great pressure that formed it.

Basalt is crystallized lava that once poured red hot from a volcano.

Volcano

# Landscapes

The surface of the Earth and the landscapes we see around us are slowly changing all the time. Rain, sun, wind and frost constantly break down rocks. Great mountain ranges are worn down to rolling plains in millions of years. Solid rock is ground into mud. We call this breaking up and wearing away *erosion*.

Running water is the most important force in changing the land. Rain washes soil down hillsides and sinks into the ground, to appear elsewhere as streams and rivers. The rivers cut into their banks and beds and carry stones and mud down to the sea. Sometimes streams travel underground and hollow out caves.

Erosion is at work in deserts, too. There the wind piles up loose sand in huge, shifting dunes. Wind-blown sand blasts exposed rocks, polishing and carving them into strange shapes.

But erosion does more than just alter the landscape. It makes soil. Soil is just surface rock that has been broken down and mixed with decayed plants.

**Some of the strangest effects of erosion can be seen in deserts, where sharp grains of sand are blown by the wind. The sand is blasted against fixed rocks, smoothing them into fantastic shapes.**

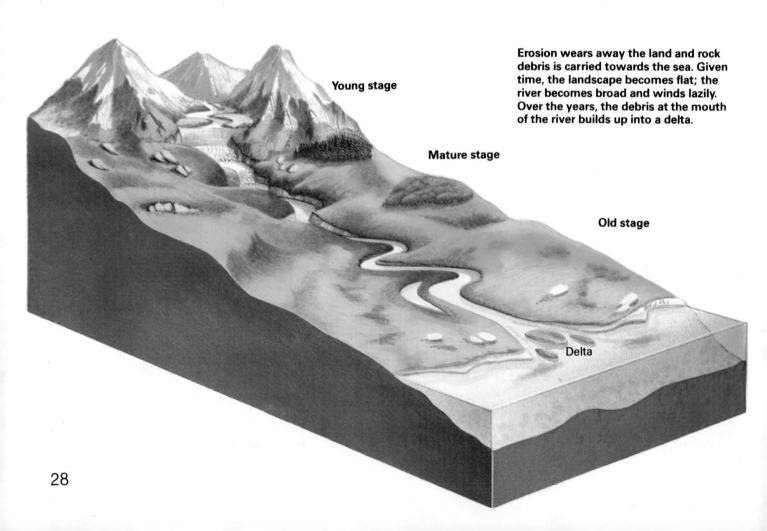

Young stage

Mature stage

Old stage

**Erosion wears away the land and rock debris is carried towards the sea. Given time, the landscape becomes flat; the river becomes broad and winds lazily. Over the years, the debris at the mouth of the river builds up into a delta.**

Delta

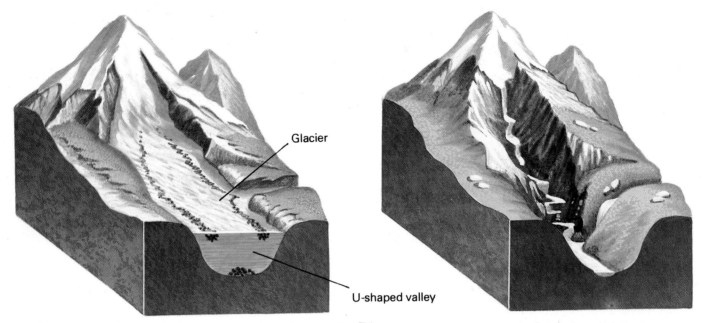

Glacier

U-shaped valley

## Frost Shattering

In cold places, ice is important in shaping the landscape. Water running down the mountain slopes seeps into cracks in the rock. As the water freezes, it expands with enough force to split the rocks apart. (The same thing happens when water freezes in our house pipes and bursts them.) Millions of tons of frost-worn rock litter mountain slopes.

## Rivers of Ice

High in the mountains, snowfalls build up into solid ice, sometimes hundreds of yards thick. These are glaciers, great rivers of ice which slowly make their way down a valley at a rate of about a yard a day. Glaciers are very powerful. As they move along, they pick up boulders and debris and carve away the valley floor and sides.

After the Ice Ages, the land had been changed by glaciers. Great U-shaped valleys had been formed. Vast areas of land were covered with boulders, sand and clay that had been moved about by the rivers of ice.

## The Sea Versus the Land

Another form of erosion is by the sea. Around our coasts, the sea is constantly eating away at the land. In some places it cuts steep rock cliffs. In others it carries sand and pebbles along the coast to dump them on gently sloping shores. So are born sand and pebble beaches. Sometimes the sand is dumped at the entrance to a bay, where it gradually forms a bar or spit which may cut the bay off from the sea.

**During the Ice Ages, rocks, firmly embedded in the glacier, turned the glacier into a giant 'file' which wore down the surrounding rocks into a deep valley.**

**When the Ice Ages passed, the glacier was replaced by a mountain river.**

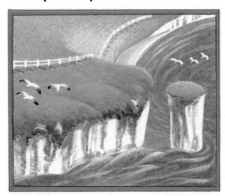

**Pounding waves wear away the land into steep cliffs.**

**Sandy beaches are built up from tiny fragments worn away from cliffs and rocky coasts.**

# A World of Atoms

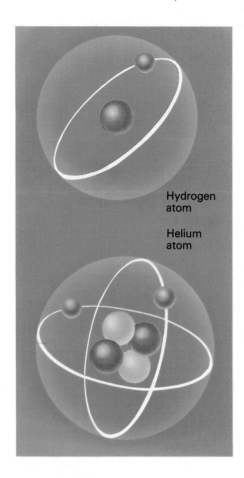

Hydrogen
atom

Helium
atom

Everything in the universe is made up of atoms. Rocks, water, the air we breathe, plants, animals and people all contain millions upon millions of these tiny, invisible particles. About 30 million atoms placed side by side would stretch across the head of a pin.

But even the tiny atom is made up of even smaller pieces. The simplest atom is that of the light gas hydrogen. At the centre of the hydrogen atom is a tiny solid body called a *proton*. Around this spins an even smaller *electron*. The whirling electron makes billions of trips around the proton in a millionth of a second. Protons have a positive electric charge, electrons have a negative charge.

Other atoms are much more complicated than the hydrogen atom. An atom of another light gas, helium, has two protons, two electrons and there are two particles called *neutrons* with the protons at the centre or *nucleus* of the atom. Neutrons have no electrical charge. The most complicated atom is the uranium atom. It has 92 electrons, 92 protons and 146 neutrons.

Atoms are so small that if an atom were the size of a finger nail, then your hand would be big enough to grasp the Earth.

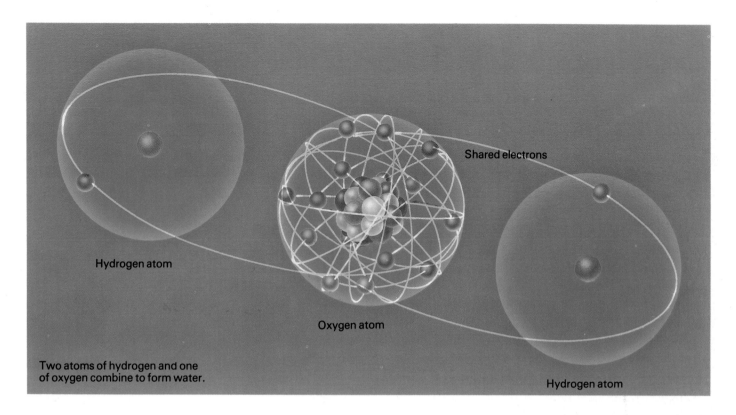

Hydrogen atom

Shared electrons

Oxygen atom

Hydrogen atom

Two atoms of hydrogen and one of oxygen combine to form water.

## The Elements

All the millions of substances in the world are made up from about a hundred simple substances called *elements*. Gold, silver and copper are elements; so are the gases hydrogen and oxygen. The atoms of different elements often join up to make different substances called *compounds*. The salt you put on your food is made up of atoms of the elements sodium and chlorine. Two atoms of hydrogen and one atom of oxygen join to make a *molecule* of water. A molecule is the smallest portion of a substance that can exist.

Every molecule of a substance is made up of the same number of atoms, joined together in exactly the same pattern. The main difference between the atoms of one element and those of another is in the number of protons in the nucleus. For example, every atom of aluminum has 13 protons; every atom of lead has 82. If an atom gains or loses protons, it becomes an atom of another element. The number of protons is called the *atomic number* of the element.

Although there are about a hundred elements, nearly all of the Earth's crust, the atmosphere and the oceans is made up of only eight elements. These are oxygen, the most common, aluminum, silicon, iron, calcium, potassium, sodium and magnesium. In the whole universe, there is more hydrogen than any other element. This is because the stars are made of hydrogen.

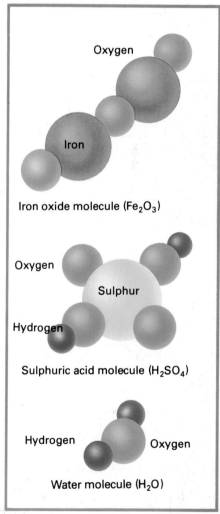

Oxygen

Iron

Iron oxide molecule ($Fe_2O_3$)

Oxygen

Sulphur

Hydrogen

Sulphuric acid molecule ($H_2SO_4$)

Hydrogen

Oxygen

Water molecule ($H_2O$)

# Solids, Liquids and Gases

When scientists talk about matter, they mean everything in the world. And all matter can be divided into three groups: solids such as iron or wood, liquids such as water and oil, and gases such as air or steam. Ice is solid water. When ice is heated it melts to become liquid water. When the liquid is heated to 212°F it boils and becomes a gas – steam.

Solids tend to resist being pulled or pushed out of shape; they usually keep the same size and shape, no matter where they are. Liquids have no shape of their own. They take the shape of any container they are poured into. Gases do not keep either their shape or their size. They expand to completely fill anything they are in.

The reason why different materials behave in different ways is because of the tiny atoms that make them up. Iron is different from gold because it is made up of a different kind of atom. The way in which the atoms are packed together decides whether a substance is a solid, a liquid or a gas.

Water is a strange liquid. It is one of the very few things that grows bigger (expands) when it freezes. That is why huge icebergs float, even though most of the bulk is under the surface.

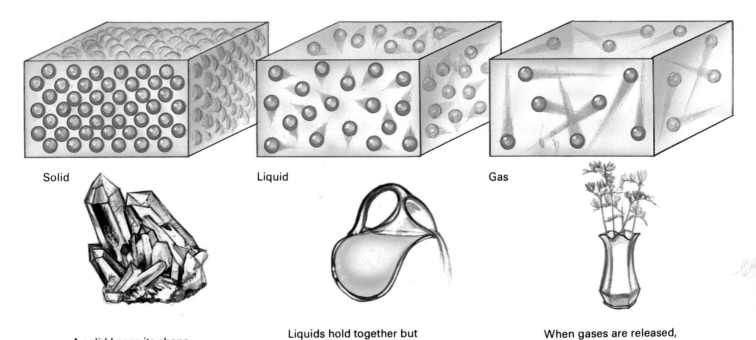

Solid

Liquid

Gas

A solid keeps its shape

Liquids hold together but take the shape of a container

When gases are released, they expand for ever

How solid something is depends on how closely packed the atoms in it are. In a solid, the atoms are close together and fixed in position. This is why it is difficult for a solid to change its shape or its size. In a liquid, the atoms are less tightly packed and can move about a bit.

When a solid is heated, the atoms in it move more and more until they form a liquid. They move apart but do not escape from each other completely. If we go on heating a liquid, the particles in it move faster and faster. After a while they move so quickly that they escape

from the surface of the liquid and become a gas. This is called *evaporation*. When the liquid gets hotter still, the particles escape so quickly that the liquid bubbles. This is called *boiling*. When water boils it turns into the invisible gas, steam.

## Solids

Solids are solid because of the way their atoms and molecules are arranged. Ice, water and steam are all the same from a chemical point of view – they all contain the same kind of atoms and molecules. The difference lies in the movement of the molecules. In ice, the molecules are held tightly in a definite pattern – what is called a *crystal lattice* – by strong forces between the neighboring molecules. Snow is a mass of beautiful ice crystals, each six-sided and no two of them are alike (see above). Although molecules in ice do not move about, they still vibrate a little.

## Liquids

Some substances such as water, oil and the metal mercury are liquid at ordinary room temperature. A liquid is similar to a gas because its atoms and molecules are not fixed together in any particular way. But it is also similar to a solid because it has a definite volume.

The molecules of a liquid are often attracted to the molecules of other substances. And this attraction is greater than the attraction between neighboring molecules of the liquid. This is why liquids will rise up a narrow tube. It is called *capillary action*. Water also has a 'skin' called *surface tension*. This is why the pond-skater (above) can walk on the surface.

## Gases

Gases behave quite differently from solids and liquids. They are the lightest and most movable form of matter. Any substance on Earth can be turned into a gas if it is heated above its boiling point. Iron becomes a gas if it is heated to about 5,250°F. The temperature of the Sun is so high that all the matter in it is in the form of gas.

Every gas consists of molecules flying about and colliding with each other. The pressure of a mass doubles if its volume is halved (above). But the temperature must remain constant.

33

# Heat – Molecules in Motion

Center of the Sun
27,000,000°F

Surface of the Sun
9,932°F

Iron Melts
2,804°F

Sunlit side of Mercury
707°F

Paper catches fire
543°F

Water boils
212°F

Hottest shade temperature
on Earth 136°F

Water freezes
32°F

Coldest temperature on Earth
−127°F

Air becomes liquid at about
−330°F

Absolute Zero
−460°F

What is heat? Scientists say it is a form of energy – the energy of moving atoms and molecules, that everything is made up of. Atoms and molecules are always on the move, and the heat of anything is simply a measure of how fast they are moving. The faster they move, the hotter a body is.

We measure heat with *thermometers*. Water boils when the thermometer shows 212 on the Fahrenheit scale (212°F); it freezes at 32°F. Our bodies use the food we eat as fuel to keep our temperature at about 98.6°F. But temperature and heat are not quite the same thing. If we put two pots on the stove, one full of water, the other with very little water, it will take much longer for the full pot to boil. This means that much more *heat* has to be put into the full pot to get both pots to 212°F.

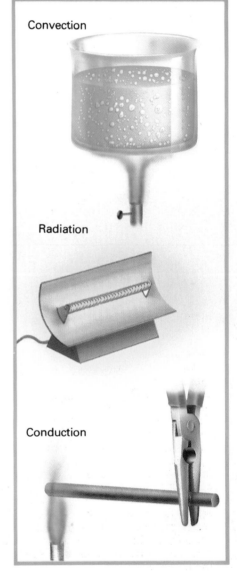

Convection

Radiation

Conduction

**Heat passes from one place to another in three different ways. They are called *convection*, *radiation* and *conduction*.**

**Convection carries heat by circulating it in *convection currents*. A room heater, for example, warms the air around it. This heated air expands and rises and is replaced by cooler air. Then the new cooler air is heated and rises. This means that a constant current of air carries heat all over the room. Convection currents also occur in liquids (right).**

**With radiation, heat travels through empty space. When something gets hot, its moving atoms and molecules make invisible waves of radiant energy. These waves are also called *infra-red rays*. The heat that reaches us from the Sun has traveled by radiation rays at the speed of light. Heat waves and light waves are exactly the same except for their different wavelengths.**

**Conduction is the movement of heat through a material or from one body to another if the bodies are touching. If we place one end of a metal spoon in boiling water, the handle of the spoon soon becomes too hot to hold. The heat from the water has traveled up the spoon by conduction. Some materials such as metals are good conductors – they conduct heat easily; other substances such as wood are bad conductors.**

**Left: An example of the tremendous range of temperatures experienced in the universe.**

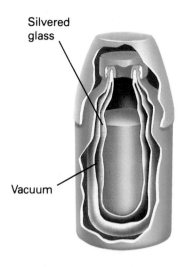

Silvered glass

Vacuum

Vacuum flasks keep things hot or cold for a long time. There is a vacuum between the double walls of the flask to stop heat loss by conduction. The container is also silvered to help stop loss of heat by radiation.

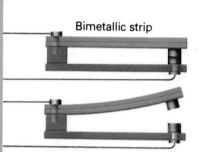

Bimetallic strip

Thermostats make use of the fact that some substances expand and contract more than others when they are heated. They are used in heating systems, cookers and irons to switch the heat on when the temperature falls too low, and to switch it off again when the required temperature is reached. They are also used in refrigerators.

Many thermostats have a *bimetallic strip* made of two different metals such as brass and iron fastened together. As the temperature rises, the brass expands more than the iron. This makes the strip bend upwards. The electrical contact is broken and the heating current stops flowing. As the bimetallic strip cools down again the brass contracts until the two metals are the same size once more. The contact is made and the heating current flows once more.

## Hot and Cold

Steel begins to melt at a temperature of about 2,700°F, but this is very chilly compared to the heat of the Sun. Inside the Sun the temperature is about 27 million degrees. Other stars are much hotter still. There appears to be no limit to how hot it can get.

Cold is different. The coldest place on Earth is a chilly −127°F (159 degrees below freezing). But the thermometer would have to drop to −297°F before the air started to freeze. At −460°F, *absolute zero* is reached. At this temperature everything would be frozen solid. Nothing would move. Even the atoms would stop moving. But it is not possible to reach this temperature.

Nearly everything grows bigger when it is heated. If you place a thermometer in hot water, the atoms in the mercury move faster and faster and take up more space. The mercury expands and moves up the thermometer's stem.

A fire-fighter wears a special suit so that he can walk through flames and remain unharmed. The suit is made of asbestos, a substance that is a very poor conductor of heat. The silvery surface of the suit reflects heat away.

# Chemical Reactions

Chemistry is the study of substances. It looks at what they are made of and how they split up or join together with other substances. Everything around us is made of chemicals. The water we drink, the salt and sugar we eat are chemicals. So are the proteins that make up most of all plants and animals.

There are just over a hundred basic chemicals called *elements*. Everything is made up of these. Iron, oxygen, carbon, gold and silver are all elements. And elements are made up of tiny atoms. Each element has its own kind of atom that is different from the atoms of all the other elements. When the atoms of two or more different elements join together, they form a chemical *compound*. Water is a compound of the elements of hydrogen and oxygen. Two atoms of hydrogen join with one atom of oxygen to make a *molecule* of water.

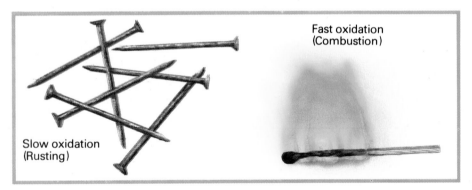

Fast oxidation
(Combustion)

Slow oxidation
(Rusting)

Some compounds are very complicated. Each molecule of sugar, for example, contains 22 hydrogen atoms, 11 oxygen atoms and 12 carbon atoms. Sugar, starch and alcohol all have molecules that contain hydrogen, oxygen and carbon, but in different proportions. It is the different proportions that make the three substances different.

Chemists use symbols to name substances. Chemical formulas show the elements that make up the substances. The symbol for the element hydrogen is H; for oxygen it is O. The formula for water is $H_2O$.

Vast quantities of chemicals are used in the modern world. Soaps and detergents, dyes and acids, polishes, artificial fibers and explosives – all these things and thousands more are products of the vast chemical industry.

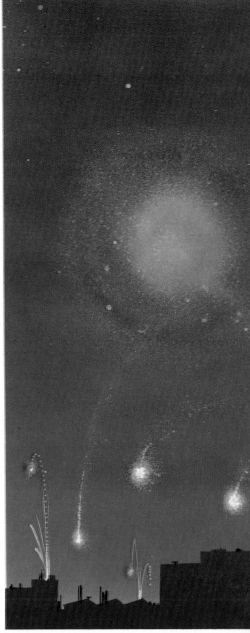

When something burns very quickly indeed, we say it 'explodes'. The exploding force in fireworks (above) comes from the rapid burning of gunpowder. Gunpowder is a mixture of sulphur, saltpeter and charcoal. When an explosion happens, a quite small quantity of explosive such as gunpowder burns in a flash and turns into a lage amount of hot gas. It is this expanding gas that sends rockets into the air. The brilliant colors of fireworks come from metallic salts that are added. Calcium salts give a red color; sodium, yellow; barium, green; and copper, blue and green.

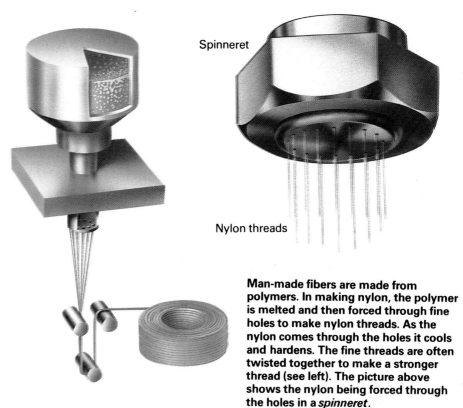

Spinneret

Nylon threads

Man-made fibers are made from polymers. In making nylon, the polymer is melted and then forced through fine holes to make nylon threads. As the nylon comes through the holes it cools and hardens. The fine threads are often twisted together to make a stronger thread (see left). The picture above shows the nylon being forced through the holes in a *spinneret*.

Substances that are made up of long chains of carbon atoms are often called *polymers*. Cotton is a natural polymer because the fibres of cotton are made up of a polymer called *cellulose*. Cellulose is a compound with long chains of carbon atoms. Nowadays, chemists make artificial polymers. For example, molecules of the gas ethylene join together in a long chain to make polyethylene – the plastic we call Polythene (see above). There are many different plastics that rely on the joining together of carbon atoms. Plastics have countless uses.

Sodium       Hydrochloric   Sodium
hydroxide    acid           chloride        Water

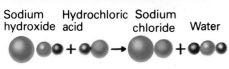

*Acids* turn litmus paper red. *Bases* turn litmus paper blue. Acids and bases neutralize each other – they cancel each other out. Bases that dissolve in water are called *alkalis*. When a base neutralizes an acid it makes salt and water only. If we take a solution of sodium hydroxide (a base) and add it to hydrochloric acid in the right quantities, the result is neutral. We are left with sodium chloride (table salt) and water (see above).

## Slow and Fast Burning

Oxygen is the most plentiful of all the elements in nature. Although it is a gas we cannot see, it accounts for about half the weight of most rocks and minerals. A fifth of the air we breathe is oxygen, and nearly all living things need it.

Oxygen is a very active chemical. It combines with many other chemical elements to make a very large number of compounds. These compounds are called *oxides*. The process in which they are made is called *oxidation*. Slow oxidation happens when iron is in damp air. This produces *iron oxide*, which we call rust. When oxygen and another element are combined rapidly, light and heat are given off. We call fast oxidation *combustion*, or burning.

# Iron and Steel

Gas outlet

Hot furnace gas

Ore, coke and limestone in

White heat

Hot air blast

Slag

Molten iron

Imagine our world without metals – no metal automobiles, coins, saucepans, tools or machines. Our civilization just couldn't exist without them. Metals are useful because they have important qualities that other substances do not have. They vary a lot in appearance and how they behave, but most of them are silvery in color and quite heavy. Many are shiny and conduct heat and electricity well. Most of them can be drawn out into wires and hammered into sheets.

But a few are different. Some are not silvery – gold and copper, for example. Others are quite light – potassium will even float on water. Mercury is a liquid at normal temperatures.

We also call mixtures of different metals 'metals'. These are *alloys* such as brass, bronze and pewter. Alloys and other metals that do not contain any iron are called *nonferrous* metals.

About a quarter of the Earth's crust under our feet is made up of metals.

Henry Bessemer invented the blast furnace. He found that if hot air was blown through molten iron, the carbon in the iron and other impurities were blown away as gases. The furnace is fed with ore and other substances. The charge melts in a fierce blast of air. It turns into molten iron, waste slag and hot gas. The waste hot gas is used to heat the air blast.

## Smelting and Refining

Metals are seldom found in the earth in their pure form. They are usually mixed up with other elements and earth materials. These mixtures are called *ores*. Separating metals from their ores and preparing them for use is called *metallurgy*.

The metal ores coming from the ground are usually just lumps of rock. The rock must pass through several stages before pure metal is obtained. Unwanted material must first be removed. This can be done by crushing, washing, heating and floating the ore in a frothy liquid. Many of the common ores such as iron ore are then heated with coke in a huge furnace. This is called *smelting*. Smelting makes a metal that is still not pure. It still has to be *refined*.

There are several ways to refine metals. Sometimes the metal is heated with substances that remove the impurities. Steel is made in this way. Other metals, such as copper, have an electric current passed through them in a solution.

When the metal has been refined it can be used in its pure state or it can be mixed with other metals to form a useful alloy. It can be shaped in a *cast*, rolled into sheets or pulled out into wires.

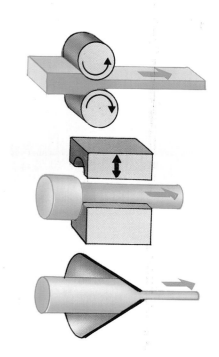

The diagrams above show three ways of shaping metals. *Rolling* (top) is a way of making sheets of metal by passing the hot metal between rollers. In *forging* (centre) the hot metal is pressed into shape between heavy blocks. In *cold drawing* (bottom) the cold metal is pulled through holes to make wire of various thicknesses.

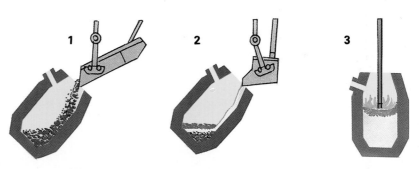

Steel is made by heating scrap steel (1) and mixing it with molten iron (2). All impurities are burned off by feeding in oxygen (3). The furnace is tilted and, after sampling (4), the molten steel is tapped off (5). Lastly, the waste slag is removed (6).

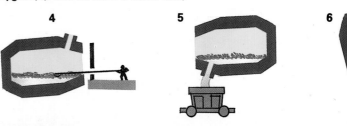

Left: Charging a steel furnace with molten iron. Steel is basically an alloy of iron and carbon, but small amounts of other metals are added during the steelmaking process to produce steels with particular properties. The addition of chromium and nickel produces stainless steel, while tungsten makes a very hard steel suitable for high-speed cutting tools.

# The Force of Gravity

An astronaut, hundreds of miles above the Earth, is weightless because he is so far from the Earth's gravitational center. He feels he is floating and not moving, but both he and his spaceship are traveling at about 18,500 miles per hour. The pull of gravity holds him in orbit around the Earth.

Everything in Earth is pulled downwards by a strange force called gravity. The pull of gravity is always towards the center of the Earth. A stone dropped from someone's hand in England falls to the ground in the same way it would fall in New Zealand on the opposite side of the world. Both stones fall towards the center of the Earth. And gravity isn't a force that happens only on Earth. Everything in the universe is attracted towards everything else. It is only because the Earth is so big and so close to us that we notice gravity here. The more massive a body, the more material in it, the greater its gravitational pull on other bodies.

The Sun is much more massive than the Earth and all the other planets put together. So the Sun's enormous gravitational pull holds all the planets in place as they circle their big parent body. It is incredible to think that our mighty Earth, speeding along at 18½ miles per second, is held in its orbit by this invisible bond. In exactly the same way, Earth's gravity holds the Moon in place in its monthly journey around us.

Gravity decreases with distance from the Earth and the same object weighs less and less.

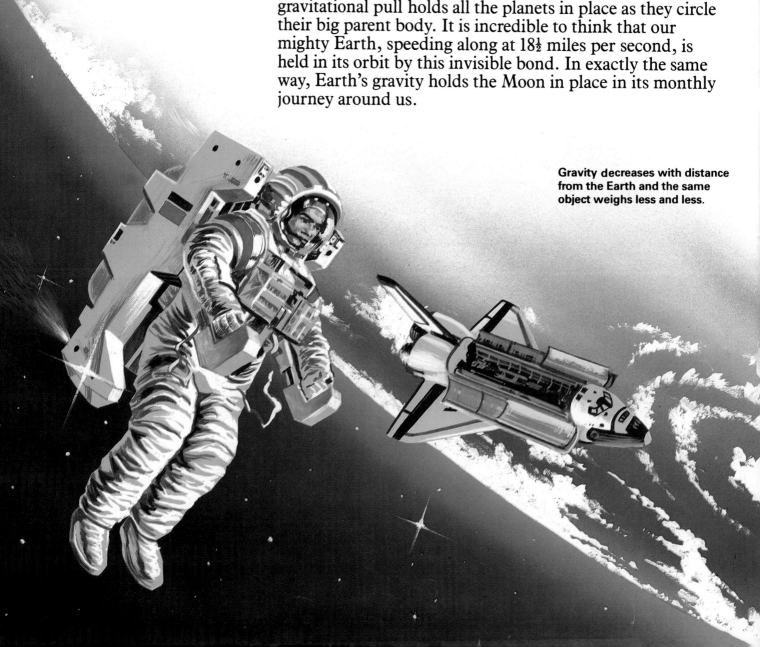

### The Tides

Although the Moon's gravity is much less than the Earth's, it still affects us. Tides are caused mainly by the gravitational pull of the Moon, and, to a lesser extent, by the pull of the Sun. When the Moon is overhead, the oceans' waters are drawn towards it, causing a bulge, which is balanced by another bulge on the opposite side of the Earth. High tides are called *spring* tides. They happen when the Earth, Moon and Sun are in a straight line (see diagram). The combined gravitational pull of the Moon and the Sun makes high tides even higher and low tides even lower. The smallest tides — called *neap* tides — happen when the pull of the Moon is at right angles to that of the Sun. Spring tides happen about twice a month, about the time of the full moon and the new moon. Neap tides happen around the first and last quarters of the moon.

Tides occur twice every 24 hours 50 minutes, the time taken for one complete orbit of the Moon around the Earth.

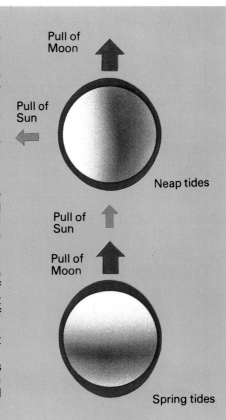

Pull of Moon

Pull of Sun

Neap tides

Pull of Sun

Pull of Moon

Spring tides

The great Italian 16th century scientist Galileo was the first to prove that all objects fall to the ground at the same speed. People knew that a cannon ball falls faster than a feather. But Galileo proved that this was because the feather was slowed down by air resistance. On the Moon, where there is no air, the cannon ball and the feather fall at exactly the same rate.

## Losing Gravity

The force of gravity grows less and less the further apart bodies are. When people go up in a spaceship the pull of Earth gravity gets less the higher they go. After a while, the Earth's pull is so small the astronauts do not notice it. They are weightless and live in a strange floating state where there is no 'up' or 'down'.

But what happens if the spaceship goes further and gets close to the Moon? Then the astronauts and their ship begin to come into the pull of the Moon's gravity. With no rockets firing, the spaceship will be pulled faster and faster towards the Moon. If the astronauts land on the Moon they find they can do things they cannot do on Earth. They can lift rocks six times as heavy. Even in their bulky spacesuits they can jump much higher. This is because the gravity of the Moon is only a sixth of Earth gravity. The Moon has only a sixth of the mass of our Earth. If an astronaut who weighed 143 pounds on Earth weighed himself on the Moon, the scales would show a weight of only 24 pounds. On the other hand, if people ever reach the giant planet Jupiter they will find things much more difficult. If you can jump a height of 3 feet on Earth, you could only jump 1 foot on Jupiter. And if it were possible to stand on the surface of the Sun, you could not even jump to the height of 1 inch!

# Magnets

A magnet is any piece of metal that will attract or pull towards itself iron, steel or a few other metals. Magnets can be of different sizes and shapes, and they can be strong or weak. The ends of magnets are called their *poles*. One end is called the *north-seeking pole* (N); the other is the *south-seeking pole* (S).

Magnets are very important. They are used every day in telephones and in the loudspeakers of television sets and radios. And they are a vital part of the big generators that make our electricity.

The magnet on the right is called a *horseshoe magnet*. If we hang chains of pins from it, each pin becomes a small magnet; each with its own north and south poles.

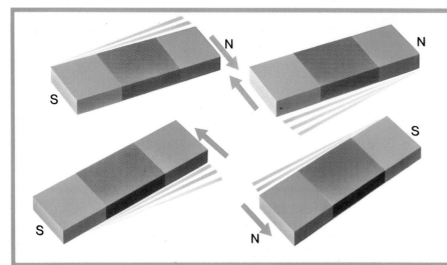

The magnets on the left are called bar magnets. If we place the south pole of one magnet near the north pole of another, the magnets will be attracted to each other. If we place the north pole of one near the north pole of another, they will push each other apart. We say that unlike poles attract and like poles repel. All magnets have more magnetic pull at their ends than at their middle.

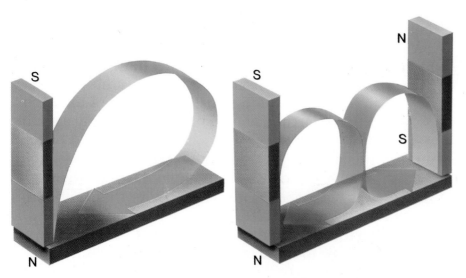

There are several ways of making magnets. One way is to stroke a permanent magnet across the metal to be magnetized, usually a piece of iron. (Soft iron is easier to magnetize than hard steel.) The iron must be stroked in one direction only, as shown in the pictures on the left.

A weak magnet can also be made by placing the iron in line with the Earth's magnetic field and hammering it. An electric current flowing in a coil around the metal will also magnetize it.

Magnets can be made to lose their magnetism by hammering them or by heating them in a flame.

# Magnetic Fields

Every magnet has an invisible *magnetic field* going through it and around it. The field around a bar magnet can be seen if we lay a sheet of paper over the magnet and sprinkle iron filings on the paper. When the paper is tapped, the iron filings will move into lines, called *lines of force*, around the magnet. Most of the lines cluster round the ends of the magnet where the magnetism is strongest.

The magnet on the right is called a *horseshoe magnet*. If we move a small compass around in the horseshoe magnet's field and note the way the compass needle points, we can draw a pattern of lines as in the picture.

The Earth has a weak magnetic field, rather like that of an enormous bar magnet. Compass needles all over the world point north and south because of the Earth's magnetism.

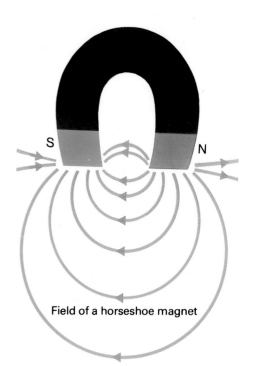

Field of a horseshoe magnet

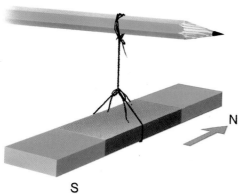

**If a bar magnet is suspended as shown above, it will always come to rest pointing in a north–south direction. And always the same end of the magnet points north. A compass (below) is really a small, lightweight magnet. It is pivoted so that it can move freely. The Earth's north pole always attracts the magnet's south pole.**

**The end of the magnet that points north is called the magnet's *north-seeking pole*.**

Compass

The Earth's magnetic field

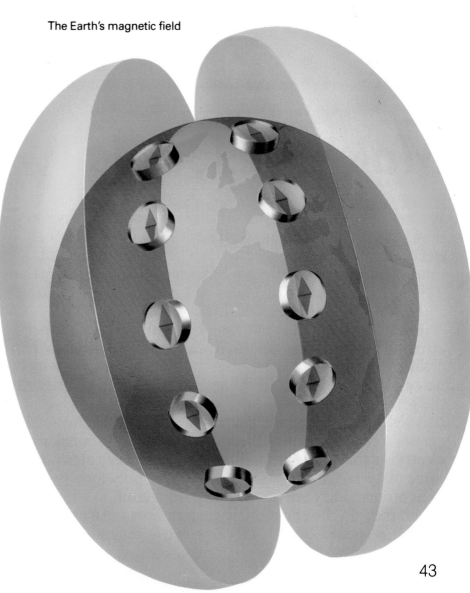

# *Making Electricity*

When ancient people saw lightning flashes in the sky, they thought the gods were angry. They did not know about electricity, but they noticed that some things seemed at times to attract other things. The ancient Greeks knew that if they rubbed a piece of amber with a woollen cloth, straw and dry leaves were attracted to it.

Today, we know that both the lightning and the amber's attraction are forms of electricity. Lightning happens when clouds store up too much electricity. Electric sparks which we call lightning shoot from the clouds to other clouds or to the ground.

An electric current is a movement, or flow, of tiny particles called *electrons*. Electrons are particles of negative electricity that circle around the center of every atom. In some materials, a few of the electrons are only loosely held to their atoms. They are free to jump from atom to atom. When they do this, an electric current flows. An electric current is started by a battery or electric generator. If the

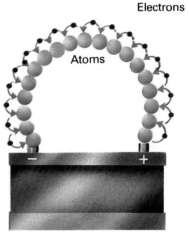

Electrons

Atoms

When the terminals of a battery are connected by a wire, an electric current flows from one terminal to the other. Most metals are good conductors of electricity – especially copper and silver. Wires are usually made of copper. The copper atoms have *free* electrons that can be pushed on to the next atom in the line. Another free electron is pushed from that atom, and so on to the other battery terminal. This is an electric current.

Electricity can be produced by separating two different metals with a solution that conducts electricity. A 'dry' cell is not really dry. It is filled with a damp chemical paste. The positive terminal is a carbon rod. The zinc container is the negative electrode.

An accumulator or battery contains cells made of lead plates in dilute sulphuric acid. Automobile batteries usually have six 2-volt cells. They are connected in series to give 12 volts.

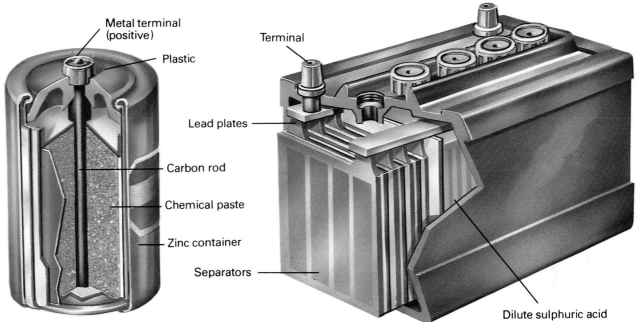

Metal terminal (positive)
Plastic
Carbon rod
Chemical paste
Zinc container
Separators

Terminal
Lead plates
Dilute sulphuric acid

terminals of a battery are connected to each end of a piece of wire, electrons are pushed from the first atom in the line to the next, and so on along the wire – all in a flash of time. Wires which carry electric current are often made of copper. Copper, like most metals, is a good *conductor* of electricity. It has lots of free electrons.

## Batteries and Generators

Batteries make electricity by chemical action. The most common kind of battery – the flashlight battery – is really a *dry cell*. When the chemicals in the cell are used up, the cell is dead and is thrown away. A battery is two or more cells working together.

An automobile battery is different. It is filled with dilute sulphuric acid in which are lead plates. When this kind of battery runs down it can be recharged by connecting it to an electric current. This makes the chemical action go backwards. The electrons are put back where they were and the battery can produce current again.

An electric generator is a machine that turns mechanical energy into electrical energy. The simplest generator is a loop of wire that is turned between the poles of a magnet. When the wire cuts the lines of force between the magnet's poles, an electric current is produced in the wire. This is the principle of the electric generator.

The diagram below shows a very simple electric generator. A loop of wire is turned between the poles of a permanent magnet. As the wire cuts the lines of magnetic force between the magnet's poles, an electric current is produced in the wire. The current is taken from the wire loop through carbon brushes that rub against metal rings. Large generators have thousands of loops of wire and produce a very large, steady current.

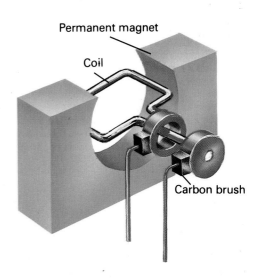

Permanent magnet
Coil
Carbon brush

# Putting Electricity to Work

When an electric current goes through the fine coiled wire *filament* inside a light bulb, the filament gets hot and glows with light. The filament is made of tungsten, a metal that does not melt easily when it is hot. The bulb has no air in it and has other gases to help stop the filament burning out.

Electricity is the most useful form of energy. It can be taken easily by cables to our homes, factories and offices and there used to produce light and heat or run machines.

The electricity we use is produced at power stations by large generators. These are machines that are turned by power from coal or oil to make electricity. Electricity flows along wires as a current. A current of electricity must have a completely unbroken path. If we could follow a current from the generator, it would travel across country through heavy overhead copper wires and along underground cables to our house. There it would go through a meter that would show how much current went through it; through fuses, to an electric light bulb. After the current has passed through the bulb and produced light, it goes all the way back through a separate wire to the generator in the power station. All this happens in a flash.

Most electricity is used to make things move. What do vacuum cleaners, food mixers and tape recorders have in common? They all have *electric motors* inside them to make things go round (see opposite page).

Some of the most powerful electric motors are used to drive electric trains. The electricity can be carried to the train's motors in different ways. Some railways have overhead wires above the track. A metal bar reaches up from the train and slides along the wire to collect the electric current. This is called a *pantograph*. Other trains get their power from a third rail placed beside the track.

## HOW AN ELECTRIC MOTOR WORKS

The diagrams show how a simple motor works. When a current flows through the coil, a magnetic field is set up. The coil then has a north pole and a south pole as shown by the 'ghost' magnet drawn as though it were inside the coil. Permanent magnets give a magnetic field in which the coil turns. Forces of attraction and repulsion between the fields make the coil turn. As the coil turns *carbon brushes* rub against separate *commutator* segments to carry current to the coil as shown. When the poles of the coil are almost in line with the poles of the permanent magnet, the brushes are almost at the end of the commutator segments (1). But the moving coil cannot stop and carries on past this point. At the same time the commutator reverses the current flowing through the coil and in doing so reverses the poles of the coil (2). (This is shown in the diagram by the ghost magnet. The end with the black dot has changed from blue to red.) Forces of attraction and repulsion between the coil and the permanent magnet keep the coil turning (3) until the commutator changes the poles again (4). In this way the coil, or motor, keeps turning.

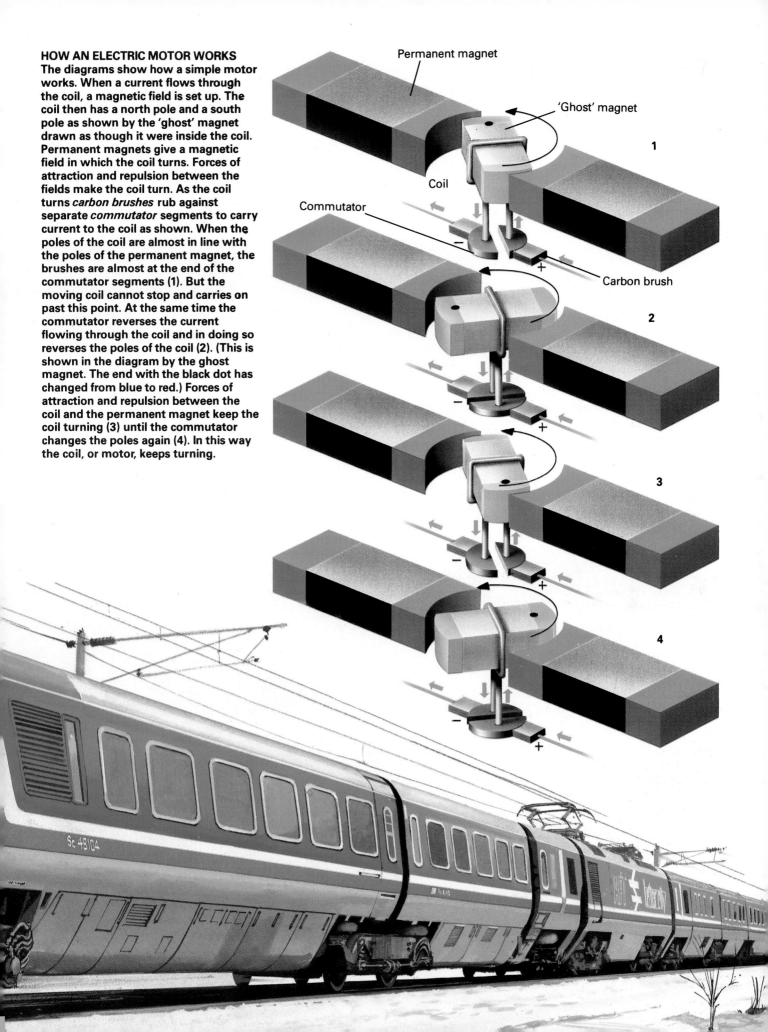

Permanent magnet

'Ghost' magnet

Coil

Commutator

Carbon brush

1

2

3

4

Sc 48104

# Man and Machines

The ancient Egyptians used the *inclined plane* to get the great stone blocks for the pyramids up to the height they needed. It was easier than lifting them straight up.

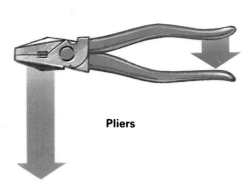

Pliers

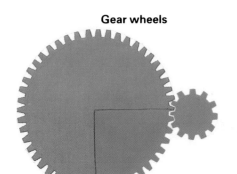

Screw

Gear wheels

A machine is something made by people to help them do jobs more easily. It may be large with masses of wheels and other moving parts like a locomotive or an automobile; or it may be very simple. A pair of scissors is a machine, and so is a screwdriver.

People began using machines centuries ago because they wanted to make their work easier. They wanted to harness power that was greater than the power of their own muscles or the muscles of animals.

Nowadays, machines are essential to everything we do. Industry uses giant machines; we use smaller machines such as washing machines, mixers and refrigerators in our homes. In fact, we depend on machines so much that a serious breakdown of machines at a power plant can cut off light, heat, transport and industrial power generally.

## SIMPLE MACHINES

All the things in the pictures above are simple machines. The lever is a simple machine. There are lots of different kinds. Scissors are levers, so are nutcrackers. Pliers are levers. Because they are pivoted near one end, a small amount of pressure on the handle end gives a lot of pressure at the other end.

The spiral thread of a screw is a kind of inclined plane. As the screw is turned, the thread pulls it into the wall.

Gears are machines that change the speed of wheels and help to do work. If the small wheel with 12 teeth turns once, the big wheel with 48 teeth makes only a quarter turn – but with four times the turning force of the small one.

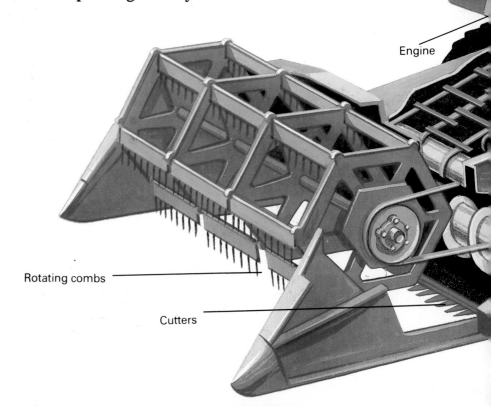

Engine

Rotating combs

Cutters

Machines give people the power to do much more work than they could with their strength alone. Suppose you wanted to shift a rock that weighed 110 pounds from one place to another. To lift it and carry it would be impossible. You would have to exert a lifting force of 110 pounds. But if you used a long board as a lever you might shift the rock by using a force of only 22 pounds. The lever, which is a simple machine, makes your work easier.

A machine can never do more work than the energy put into it. It often turns one kind of energy, such as electricity, into another kind – mechanical energy that turns wheels or moves machine parts in some way. The efficiency of a machine is the ratio between the energy it supplies and the energy put into it. The perfect machine should have an efficiency of 100 per cent, but this is impossible because every machine has friction in its moving parts. Many machines have an efficiency of only about 10 per cent. Very few can do better than 30 per cent.

For centuries people have tried to make a machine that, once started, will work for ever without needing power – a perpetual motion machine. In the machine above the magnet was supposed to pull the ball up the slope. It fell through the hole, ran down and was pulled up again – for ever. But it didn't work because of friction.

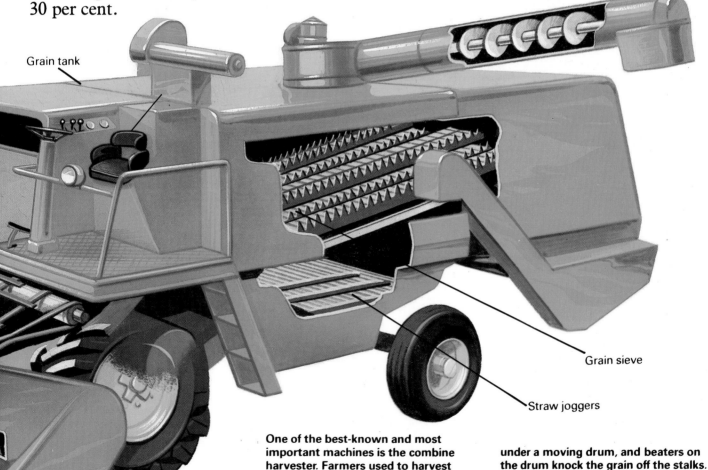

Grain tank

Grain sieve

Straw joggers

Rotating screw

One of the best-known and most important machines is the combine harvester. Farmers used to harvest wheat by hand. They cut the crop, gathered it and slowly separated the grain from the chaff. Today, a giant combine harvester does the whole job in a fraction of the time. Moving knives cut the stalks. The cut stalks are carried under a moving drum, and beaters on the drum knock the grain off the stalks. The grain is sieved to the bottom of the harvester, then carried up by a rotating screw to the storage tank. From there it is piped out into a truck. The straw stalks are shaken to the back of the harvester, where they fall to the ground.

# Computers and Robots

Computers are playing a more and more important part in all our lives, whether we realize it or not. Businesses, large and small, are using computers to keep accounts, pay salaries, keep an eye on the stock position. They are used in schools, by the police, by banks, by the armed forces, by airlines and by scientists.

The strange thing is that computers can only do a few simple tasks. They can add. They can subtract. And they can compare one number with another. Why, then, are computers so special? The answer is that they can do these three things at lightning fast speed. They can do millions of calculations in a second.

Although the computer works with numbers, the information it uses does not have to start off as numbers. It can play chess with you, guide a spacecraft, check fingerprints and draw a map of Australia. But before it begins to work on any of these tasks it turns the information into numbers. And the numbers it uses are not quite the same as ours. We use the numbers 0 to 9. All the computer needs is 0 and 1. In fact, it can only count up to 1! This is called the *binary system*.

The computer uses the binary system because it has been designed to work with electrical currents. It can recognize the difference between a flow of current and no flow of current. If there is a current it registers 1; if there is no current it registers 0.

At the heart of every computer, pocket calculator or digital watch is the silicon chip. A tiny chip only 1/5th of an inch square can be the main part of a computer. The number of microscopic transistors and other electrical parts that can be put on a chip has increased rapidly year by year. It is now possible to put more than a million of them on a single tiny chip.

Because the silicon chip is so small and cheap, computers have also become much smaller and cheaper. The home microcomputer can work very well with the household TV set.

As we press the computer's keys to give it commands, the computer translates our key commands into its own binary computer language and works on them. The result appears on the screen.

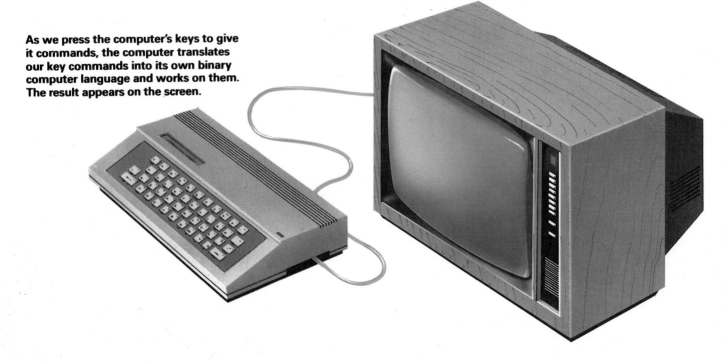

## Communicating with the Computer

To instruct a computer to do something you have to write a program. Writing a program in binary numbers would take a lot of time and effort – the binary for our 8 is 1000 and to the computer the letter T is 01010100. So a simple solution has been found. The computer itself is programmed to translate our instructions into binary. We type in our program in a language we can understand. The computer then translates our language into its own language and starts work on it.

The computer does all its calculations in its main part – called the *central processing unit* or CPU for short. It also has a memory where it stores all the information that is fed into it. It stores little bits of information in separate memory locations or 'boxes'. All we have to do is give the computer the address of any memory location and the machine will find the information in that location in a millionth of a second.

To communicate with a computer we usually type in letters and numbers as on an ordinary typewriter, but using some special computer commands.

### WHAT IS A ROBOT?

A robot is a machine that can be programmed to do different tasks. And most robots have an arm or arms that can do work for us. The robot's master is a computer.

More and more robots are working in factories all over the world. They spray paint, lift heavy loads and weld things together. And when they have been taught to do these things they usually do them better than human beings can. Switch on a robot and it will go on working 24 hours a day without stopping for a rest. It can work in places where people could not exist, and it hardly ever goes sick.

In the picture below, robots are welding cars as they move along an assembly line. Very careful programming lies behind a production system like this. A computer controls the robots so that they spot weld sections of each car without getting in each other's way.

# Sound Recording

Compared with light, sound travels very slowly. In air, sound travels at about 1,080 feet per second. Light travels about a million times faster. This means that spectators at an athletics meeting see the smoke from the starter's gun about half a second before they hear the bang. And sound needs something to travel through – something such as air or water. On the Moon there is no air. There is therefore no sound. But light and radio waves can travel through empty space, so astronauts talk to each other by radio.

Sound is very useful underwater. A ship's echo sounder sends out bursts of sound waves from under the ship. The sound travels down to the seabed and bounces back up to the ship. The echo sounder calculates the time taken for the sound to go to the bottom and back to the ship. This gives the depth of water under the ship.

To make sound, we must make something vibrate – a violin string, for example. If something vibrates, it makes the surrounding air vibrate. These air vibrations reach our ear-drums and make them vibrate too. We hear the sound.

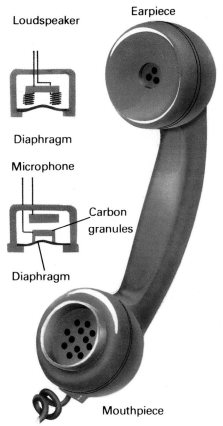

Loudspeaker
Earpiece
Diaphragm
Microphone
Carbon granules
Diaphragm
Mouthpiece

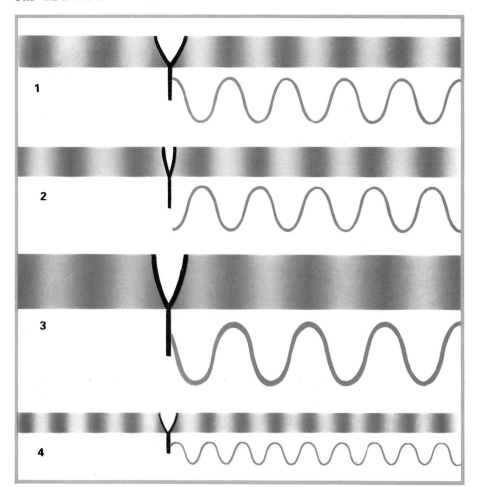

The mouthpiece of a telephone has a small microphone inside it. The sound waves from your voice make a thin diaphragm vibrate. This squeezes carbon granules in the microphone. An electric current flows through the microphone and the granules vary the strength of this current as you speak. This varying current flows through wires to the telephone exchange, from where it is sent on to the earpiece of the other telephone. There is a small loudspeaker in the earpiece which produces the sound of your voice.

Sound is made when something vibrates. If we strike a tuning fork, the prongs of the fork vibrate. As they move outward in the air, the air molecules are squeezed. A region of *compression* forms (1). When the prongs spring back, the air molecules move apart. There is a region of *rarefaction* (2). These regions of compression and rarefaction move out through the air. We call them sound waves. If the tuning fork is struck harder, the compressions are greater and the sound is louder (3). If a smaller fork is struck, the vibrating frequency is higher. A higher pitched sound is heard (4).

In a recording, sounds are recorded as a magnetic pattern on plastic tape. The tape has a coating of magnetic iron oxide on one side. The capstan rotates, pulling the tape past the heads. The erase head wipes out any existing recording on the tape by making the tape pass through a rapidly changing current. The record head causes a varying current that corresponds to the voice or music to be recorded. This current magnetizes the tape. When the tape is played back, the playback head picks up the magnetic signals from the tape. These become an electric current that is amplified and is a copy of the original sound. Most home recorders have only two heads – an erase head and a record/playback head. Many professional tape recorders, however, have three heads as shown in the diagram at the bottom of the page.

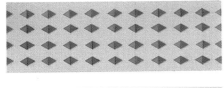

We can think of a tape as having tiny magnetic particles. Before recording, the magnetic particles are arranged as in the top picture. After recording, they become arranged in a pattern that corresponds to the sound recorded.

# Recording Sound on Disc

In 1877, Thomas Edison, the great American inventor, recited 'Mary had a little lamb' into a tube. At the end of the tube was a thin metal disc that vibrated as Edison spoke. Attached to the disc was a needle that vibrated with the disc. The vibrating needle was made to cut a wavy groove in a drum covered in tinfoil. This wavy groove was a copy of the loudness and pitch of Edison's voice. When Edison attached a horn to the tube and turned the drum again, a very scratchy voice said: 'Mary had a little lamb'. Edison had discovered how to record sound.

Today, records are made and played back by electricity. The grooves in the record are very fine and play for a long time. Sounds are picked up by a microphone which turns the sound waves into electrical waves. These electrical waves go to a sapphire needle that vibrates and makes grooves in a smooth lacquer disc. From this master disc many other records are made.

Stereo records are made with two microphones, each picking up different sounds. These separate sounds are cut into each side of the record's grooves. The result is a fuller, richer sound.

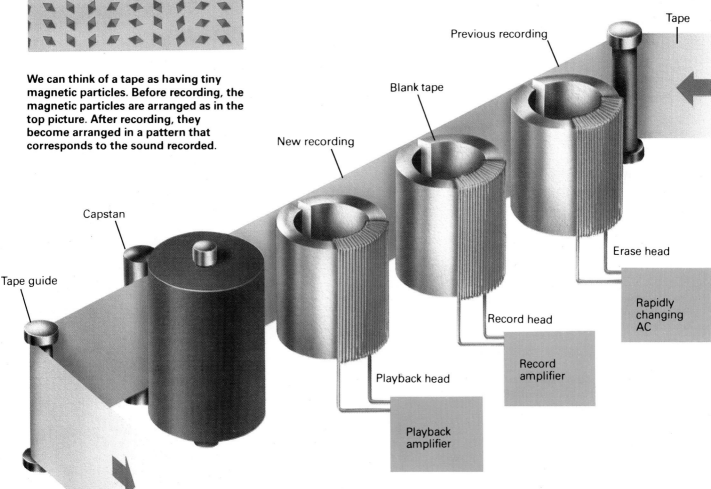

Tape

Previous recording

Blank tape

New recording

Capstan

Tape guide

Erase head

Rapidly changing AC

Record head

Record amplifier

Playback head

Playback amplifier

53

# Across the Spectrum

Cosmic rays     Gamma rays     X-ray

The Sun is constantly giving out an enormous amount of energy in the form of waves. These waves include the visible light rays that we can see. Others are infra-red rays (heat), ultra-violet rays, radio waves, X-rays and gamma rays. All these waves are forms of *electromagnetic radiation* and they all travel at the same speed – 186,000 miles per second – the speed of light.

But there is one important difference between these various waves. They all have a different *wavelength* – the distance between the start of one wave and the beginning of the next. Gamma rays are less than a billionth of an inch long, radio waves can measure several miles.

**When an atomic bomb explodes it gives off a vast quantity of heat waves and other dangerous radiation.**

**Bottom right: One of the many uses of ultra-violet radiation is in the detection of forgeries. Fluorescent substances are often present in materials such as inks, and even very slight variations show up under ultra-violet light.**

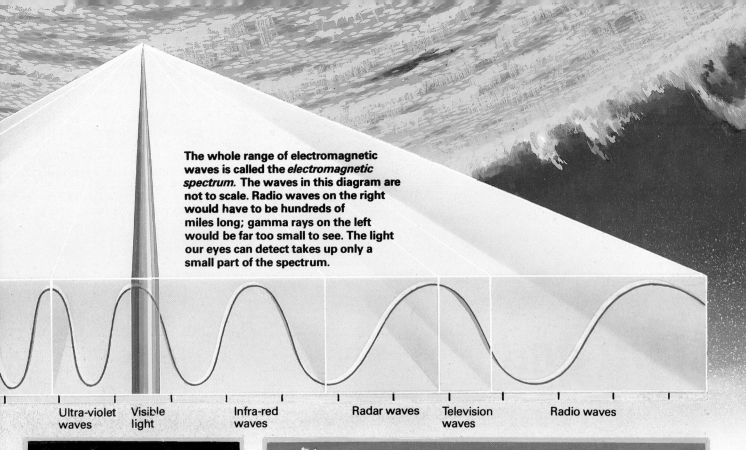

The whole range of electromagnetic waves is called the *electromagnetic spectrum.* The waves in this diagram are not to scale. Radio waves on the right would have to be hundreds of miles long; gamma rays on the left would be far too small to see. The light our eyes can detect takes up only a small part of the spectrum.

| Ultra-violet waves | Visible light | Infra-red waves | Radar waves | Television waves | Radio waves |

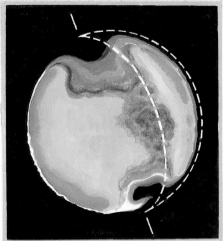

Above: An infra-red photograph of the Earth taken from space.

Left: When sunlight falls on rain or spray, we sometimes see a rainbow: the drops of water break up the Sun's light into the colors of the spectrum. The colors are always in the same order – from red to violet.

Doctors use X-rays to see inside our bodies. These very short waves pass right through some things more easily than others – bones, for example, stop the waves quite a lot. This means that when X-rays are passed through us on to a photographic plate, doctors can see broken bones and other things that are wrong inside us.

Radio telescopes are important in astronomy. Their huge dishes capture radio signals from the heavens.

Visible light from the Sun can be broken down into the colors of the rainbow – from violet at one end to red at the other. The band of radiation immediately above red is called infra-red. It has a longer wavelength than red light and cannot be seen. But we can feel it as heat. Beyond infra-red come radio waves.

Below violet light comes ultra-violet radiation. Ultra-violet rays are sent out by the Sun. They pass through our skin and reach the nerves that lie under its surface. Below ultra-violet rays come the X-rays that doctors use to see right inside us.

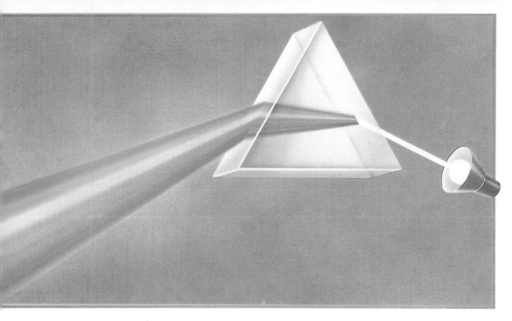

The first man to study light and tell people how it worked was Isaac Newton. In 1665, the great scientist shone a beam of light through a glass *prism,* shaped like the one on the left. He found that the light that came out of the prism had been broken up into all the colors of the rainbow. Newton had discovered that ordinary white light is made up of all the rainbow colors added together. We see a band of colors because our eyes see different wavelengths of light as different colors. Each color has its own wavelength.

# Light and Color

Without light, all life on Earth would come to an end because all the plants and trees would die. People have always realized how important light is, so they tried to find out what it was. Some thought it was made up of tiny particles, others thought it was a series of waves. Today, scientists think that light is neither completely a wave nor completely a stream of particles. It is a cross between the two. But they are still not quite sure what light really is. They do know that light waves are *electromagnetic,* just like radio waves and X-rays.

Radio waves can be miles long. Light waves are very short – about two-hundred-thousandths of an inch long. This wavelength is important because it limits the size of things we can see through a microscope. If we look at anything about the size of the wavelength of light through a powerful microscope, it is fuzzy.

**MIXING LIGHT**
The three primary colors of light are red, green and blue. Any other color can be made by mixing these colors. When red, green and blue lights are mixed, the result is white light.

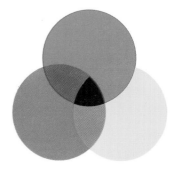

**MIXING PAINTS**
When paints, inks or dyes are mixed, the basic colors are cyan blue, magenta red and yellow. Cyan and yellow give green. If the three basic colors are mixed, the result is black.

**Invisible Light**
Different kinds of light can be seen by different animals. Most humans see all the colors from red through orange, yellow, green, blue to violet.

More than a hundred years ago, scientists tried to find out about the spectrum colors by putting a thermometer in each color coming from a prism. It was found that as the thermometer was moved from violet to the red end, the temperature increased slightly. But, even more surprising, when the thermometer was placed beyond the red, where there was no visible light, the temperature

was even hotter. There is a hot, invisible radiation just below the red. This radiation is called *infrared* radiation (*infra* means below). Some animals such as the pit viper can actually 'see' these infrared rays, which are really heat rays.

At the other end of the color spectrum are other beams of 'light' that we cannot see. They are just above the violet, so they are called 'ultraviolet'. Bees can see ultraviolet light although humans cannot.

Light travels in a straight line at a speed of about 186,000 miles per second. But there are ways of making light change direction. One way is to bounce it off the surface of something. This is called *reflection*. We can see the Moon and the other planets because they reflect the Sun's light. They have no light of their own.

## Seems Straight

Place a coin in a cup and move your head back until the coin is just no longer visible. Now, keeping your head steady, pour some water into the cup. Hey presto! The coin appears. This magic is caused by *refraction*. Refraction causes a light beam to bend as it passes from one substance to another. When the light beam from the coin leaves the water and enters the air, it bends so that you can see the coin. The coin looks as though it is in the position at the end of the dotted line in the diagram below. The amount by which light is refracted depends on two things: the angle at which the light beam strikes the second material, and the speed at which the light is traveling. If the light beam goes straight from one substance to another at right angles there is no refraction. If you put the coin in the cup and look straight down on it as you pour in the water, the coin doesn't appear to move. Light travels at its fastest in a vacuum (empty space). In air it travels almost as fast. But in water and glass, light slows down. In fact, in going through some kinds of glass, light travels at only about half its speed in a vacuum 186,000 miles per second.

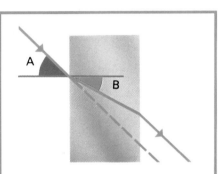

When light passes through a piece of glass it is *refracted*. Angle A is called the *angle of incidence*. Angle B is the *angle of refraction* (the angle by which the beam bends).

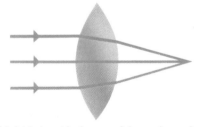

Light is bent in *lenses*. A lens shaped like the one above (convex) brings light rays together at a point called the *focal point*. A magnifying glass is a *convex lens*.

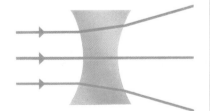

*Concave lenses* like the one above make light rays spread out. If you look through one, things look smaller. Lenses of this kind are nearly always used with other lenses.

## Very Special Light

The laser is one of the most important inventions of the 20th century. A laser beam is a beam of very pure light. We have seen that ordinary light is made up of all the colors of the rainbow. Each color has a different wavelength. A laser beam has waves that are all the same. The waves rise and fall in step. This makes the laser's narrow beam very powerful. Because of the laser's power and accuracy, it is being used for more and more tasks. It can drill a tiny hole in a diamond, slice through steel plates, help in delicate surgical operations and carry thousands of telephone messages through fine fibers of glass.

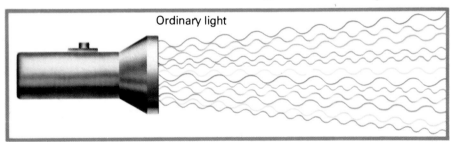

Ordinary light

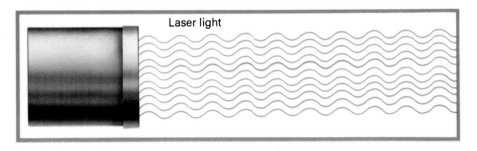

Laser light

The camera on the left is a 35 mm single-lens reflex. A system of mirrors and lenses allows the person taking the picture to view the subject through the camera's lens. The lens of this camera can be unscrewed and replaced by another type of lens such as one for taking wide-angle views or a telephoto lens for taking pictures of distant objects.

# *Photography*

The word 'photography' means 'writing or drawing with light'. To take pictures we need two things – light and some material that is sensitive to light.

To take a black and white picture with a simple camera, we load a film into the camera. The film is usually a roll of plastic. The plastic is coated with a thin layer of substance that changes when light falls on it. The roll of plastic is stretched between two spools inside the camera. By pulling a lever or turning a key we wind an unexposed part of the film into the right position in the dark of the camera (see above). We are ready to take a picture. Just point the camera towards the subject and press the button. Pressing the button works a shutter that opens and closes very quickly to let just the right amount of light into the camera. The light is focused onto the plastic film by a glass lens.

### Processing the Film

But when we have taken all the pictures possible on the roll of film, we still cannot see them. The film is now taken from its container in a dark room with perhaps only a dark red light. It is 'developed' by being placed in

Today's cameras often have several glass lenses (above). These are called compound lenses and they give a sharper image than a single lens. Most cameras have a focusing device. This makes the lens go nearer or further away from the object being photographed. We 'focus' on the object to get a clear, sharp picture by screwing the lens in or out. To take sharp pictures of close-up objects, the lens is further away from the film. When we focus on distant objects, the lens is closer to the film.

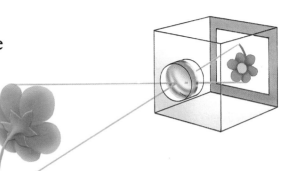

A simple way of capturing and controlling light is to use a pinhole camera. This is just a light-proof box with a tiny hole at one end. Light from the flower above goes through the hole and forms an upside down image at the back of the box. But this image is rather dim, so we need a lens to make it sharper. We also need a shutter and lens diaphragm to control the amount of light that reaches the film at the back.

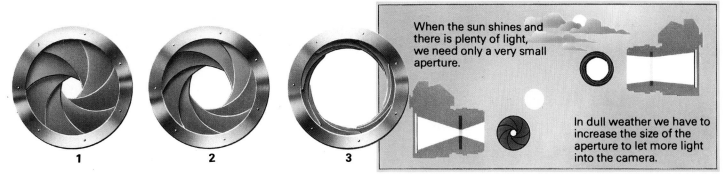

When the sun shines and there is plenty of light, we need only a very small aperture.

In dull weather we have to increase the size of the aperture to let more light into the camera.

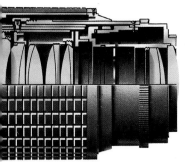

## APERTURES

The aperture is the size of the opening of the diaphragm. In most cameras the size of the opening can be varied. The diaphragm can be closed to allow only a small amount of light into the camera (1). When it is fully open (3), a large amount of light passes through. Apertures are measured in 'f-numbers' – a low f-number such as 2 means that the aperture is very wide. A high f-number such as 16 means a small aperture. There are advantages in using a small aperture. An object far away from the camera will be quite sharp and clear – it will be in 'focus'. And so will an object quite close to the camera. If a large aperture is used, objects in the foreground and background will be fuzzy.

Once the photographs have been taken, they are processed in a darkroom – only a small red light called a safelight can be used. Three trays are usually needed. The first one contains the developer, the second the 'stop' liquid that stops the action of the developer, and the third the fixer chemical that stops the film being sensitive to light. The object on the left is a masking frame to hold the paper while printing. On the far left is an enlarger, used for making prints larger than the negative.

a special liquid that changes the chemicals on the film's surface. The pictures begin to appear. The film is now a 'negative' – dark areas in the picture are light, light areas dark.

To print the film, light is shone through the negative onto printing paper. Chemicals in the printing paper make our pictures appear as the camera saw them.

# Radio and Television

## Inside the Camera

The job of a television camera is to turn the image it sees into electrical signals that can be transmitted. In color TV, the camera usually has three separate tubes inside it. These tubes split up the light from the image into three parts – a red part, a green part and a blue part. This splitting up is done by special mirrors called *dichroic mirrors*. You can see how this works in the diagram on the opposite page.

Each light color goes through a special tube in the camera. These tubes make a pattern of electric charge as light falls on them. A beam of electrons in each tube moves quickly over the pattern of electric charge, going from left to right and top to bottom. This is called scanning. It makes a stream of electric signals, each signal telling how bright or dark a tiny part of the whole picture is.

## From the Studio to Your Home

Several cameras are used to give different views of whatever is being televised. The cameras turn the image of what they see into electrical signals. These signals are fed to a control room where the program director sits in front of a row of TV screens. Each screen shows the picture from one of the

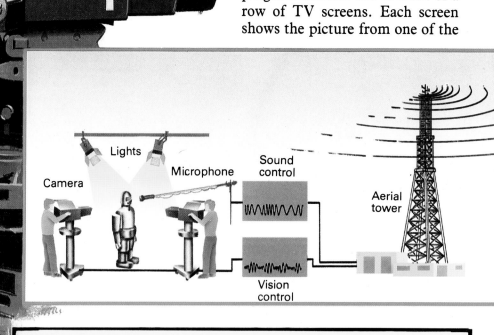

Lights

Microphone

Camera

Sound control

Vision control

Aerial tower

## Radio Waves

Radio waves are quite invisible, but we know that they can be of many different wavelengths. (The wavelength is the distance between the top of one hump in the wave and the top of the next hump.) Some are very short – only a few inches long, others can be several miles long. Unlike sound waves, radio waves do not need air to travel through. We can talk to astronauts on the Moon because the radio waves travel quite easily through empty space. When a radio wave hits the aerial of your radio, it sets up a tiny electric current. If the set is tuned to the wavelength of the radio wave, the circuits in the set get rid of the carrier wave and send the signals of a person's voice or music to your loudspeaker.

cameras. The director chooses which picture he wants at a particular time, and the signals from this picture are fed to the transmitter. The transmitter may be quite a long way away from the studio.

Sound in the studio is picked up by microphones – again there can be several. The microphones turn the sound into electrical signals which go to a sound mixing position in the control room. There the sounds are selected or mixed as required before being fed to the transmitter.

The vision and sound signals are carried on radio waves sent out from the aerial at the top of the big tower. These radio waves are picked up by a receiving aerial and go into your TV set. The set turns the signals back into pictures and sound that you can see and hear.

## Inside Your TV Set

When the picture signals go into your TV set they are removed from the radio wave that has carried them and go into a cathode ray tube. It is the front of this tube you look at when you watch TV. An electron gun for each color shoots out electrons which strike the back of your screen. This is covered with different types of tiny phosphor dots that glow, blue, red or green when they are hit. The beams go through a perforated shadow mask that ensures that each beam strikes only one type of dot. As the beams travel over the screen very quickly, our eyes see a picture like that in the TV studio.

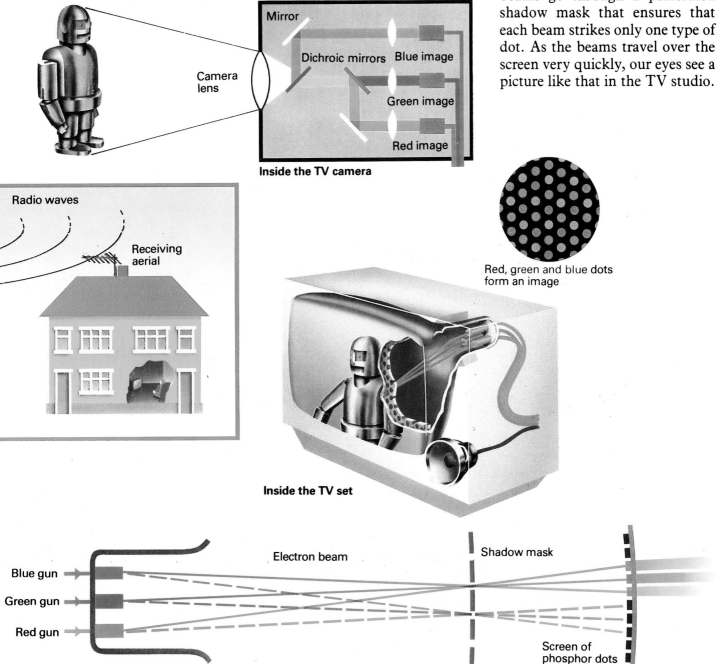

**Inside the TV camera**

Mirror
Dichroic mirrors — Blue image
Camera lens
Green image
Red image

Radio waves
Receiving aerial

Red, green and blue dots form an image

**Inside the TV set**

**Inside the TV tube**

Blue gun
Green gun
Red gun
Electron beam
Shadow mask
Screen of phosphor dots

61

# Heat Engines

An engine is any machine that takes energy from heat, water or wind and makes this energy do useful work. The windmill and the waterwheel are simple kinds of engine, and people are still trying to make these engines more efficient. Steam engines took the place of the windmill and the waterwheel, and today we have gasoline engines, diesel engines, jet engines and turbines.

A steam engine is a *heat engine*. Heat from burning coal, oil or gas is used to turn water in a boiler into steam. When water boils to become steam it expands to about 1,700 times its size. Steam engines use the energy of expanding steam to drive wheels or do other work.

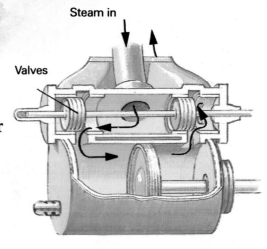

In a simple steam engine a piston slides to and fro inside a hollow cylinder. A system of valves allows steam into the cylinder at one end, then at the other, driving the piston back and forth.

## How a Steam Engine Works

In a steam engine, the expanding steam pushes a piston to and fro inside a tube called a cylinder. The piston is attached to a piston rod that moves in and out with the piston. The piston rod is attached to another, longer rod called the connecting rod, which is joined to a driving wheel and makes it go round.

Various systems of valves allow the steam to shoot into the cylinder so that it drives the piston first one way, then the other. This kind of engine is called a *reciprocating* engine. At the beginning of the 1900s the reciprocating engine was the chief source of power. It ran locomotives, ships, factory machines and even automobiles. Today it has almost vanished because more efficient kinds of engines have been invented.

Another kind of steam engine is the *steam turbine*. A turbine is a large wheel with dozens of blades round it. A powerful jet of steam is made to hit the blades and cause the turbine to spin. Spinning turbines can be used to make electricity or drive a ship's propellers.

## Internal Combustion

The motor car engine is called an *internal combustion* engine because the fuel is burned inside the engine. (In the steam engine the fuel is burned outside, away from the moving parts.) The internal combustion engine is more efficient than the steam engine. It gives more power for the energy put into it.

## How it Works

The fuel – gasoline or diesel oil – burns in a hollow cylinder. As the fuel turns to gas, it expands and

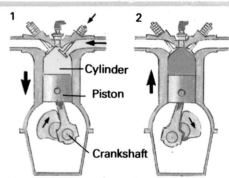

Gasoline engines do not give power on every stroke of the piston. The diagrams above show what happens. At (1) an inlet valve opens and a mixture of gasoline and air is sucked into the cylinder. Then the valve closes and as the piston goes up it squeezes the fuel

pushes a tight-fitting piston down the cylinder. When the piston is pushed down, it turns a *crank-shaft*. The crankshaft is made to turn the automobile's wheels.

## The Wankel Engine

The Wankel engine is an internal combustion engine like the ordinary gasoline engine, but there are no pistons moving in cylinders. Instead of pistons, the Wankel has a central rotor in the shape of a triangle with curved sides. As the rotor goes round the combustion chamber, the engine goes through the intake, compression, power and exhaust stages of an ordinary engine.

## Jet Engines

There are several kinds of jet engine. The first to be invented, and still very much used, is the *turbojet*, pictured in the diagram below. It works by eating up air in enormous quantities – it needs the oxygen in the air to make the fuel burn properly. As air is sucked in at the front by a series of fast-spinning blades, it is *compressed* – squeezed into a small space. The air has to be compressed to give a lot of oxygen in the combustion chamber.

Special paraffin is sprayed into the combustion chamber. This fuel spray goes on all the time the engine is working. It is first ignited

The engine above is a *turboprop*. In this engine the turbojet is used to turn a propeller which drives the aircraft forward. The turboprop is used on smaller aircraft, in which it is most efficient.

The *turbofan* engine pictured below is the one that powers most of today's big airliners. These engines have huge fans at the front to push enormous quantities of air back into the compressor. The air is divided into two streams; one goes through the combustion chamber, the other flows past the engine itself. The two streams combine at the back to give greater thrust.

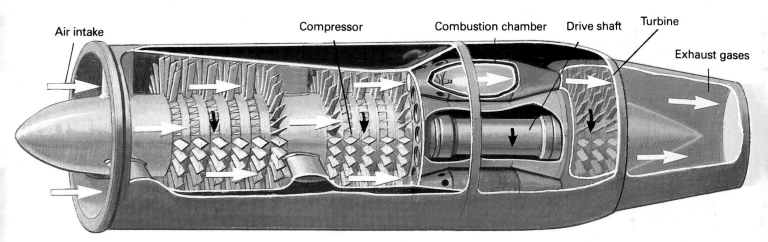

Air intake    Compressor    Combustion chamber    Drive shaft    Turbine    Exhaust gases

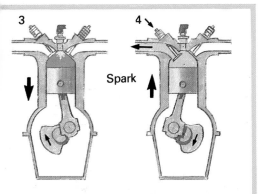

into the top of the cylinder (2). A spark takes place and the fuel mixture burns with great force and pushes the piston down the cylinder (3). Then an exhaust valve opens and the burnt gases are pushed out (4).

by electric sparks. As the fuel burns, its temperature rises to well over 1800°F. The hot gases expand and shoot out backwards into the atmosphere. This powerful stream of hot gas shooting out pushes the aircraft forward.

The hot gases turn another wheel called a *turbine*. A shaft connects the turbine to the compressor. This means that the compressor is kept turning and squeezing more air into the engine.

Jet speeds are sometimes increased by burning extra fuel between the turbine and the outlet nozzle. This is called *afterburning*.

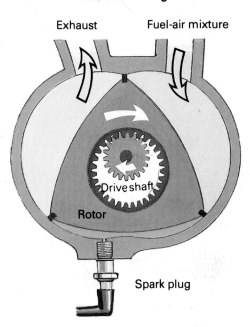

**The Wankel engine**

Exhaust    Fuel-air mixture

Drive shaft

Rotor

Spark plug

# Science in the Air

The cockpit of a modern airliner is a mass of electronics. A dazzling variety of dials and warning lights face the pilot. They keep him informed as to how all the plane's systems are working, whether he is on course and so on.

As the aircraft approaches an airport an Instrument Landing System guides it in. Ground control supplies the pilot with direction of approach. Aircraft usually start to line up with the runway about 4 to 6 miles from the airport; then they follow radio beams until they land. Landing has become more and more automatic, and many aircraft can now land without the pilot touching the controls at all. at all.

Despite all these electronic aids, aircraft still need pilots. In passenger planes, two sets of all essential equipment are carried, in case one fails. But even if both sets fail, the air crew are still there to bring the plane safely to land.

**THE ALTIMETER**
The pressure altimeter (opposite) tells the pilot how high he is above sea level. In the instrument is a sealed thin metal capsule filled with air. The pressure inside the capsule is always the same. But as the plane goes higher, the air pressure around it grows less. The air in the capsule can then push the thin metal outward. This turns a lever which works a pointer. Pipes connect the altimeter to the air outside the plane.

**AEROFOILS**
To fly, an aircraft must in some way lift itself off the ground against the pull of the Earth's gravity. This lift is produced by air flowing over the plane's wings. The wings have a special shape called an *aerofoil.* They are curved at the top and flat at the bottom. This means that the air passing over the top of the wing has to travel faster because it has further to go. So the air pressure is less above the wing than below it and the wing is lifted upward.

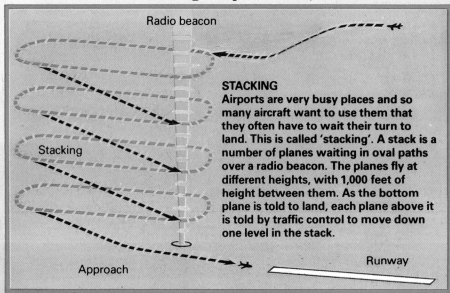

Radio beacon

Stacking

Approach

Runway

**STACKING**
Airports are very busy places and so many aircraft want to use them that they often have to wait their turn to land. This is called 'stacking'. A stack is a number of planes waiting in oval paths over a radio beacon. The planes fly at different heights, with 1,000 feet of height between them. As the bottom plane is told to land, each plane above it is told by traffic control to move down one level in the stack.

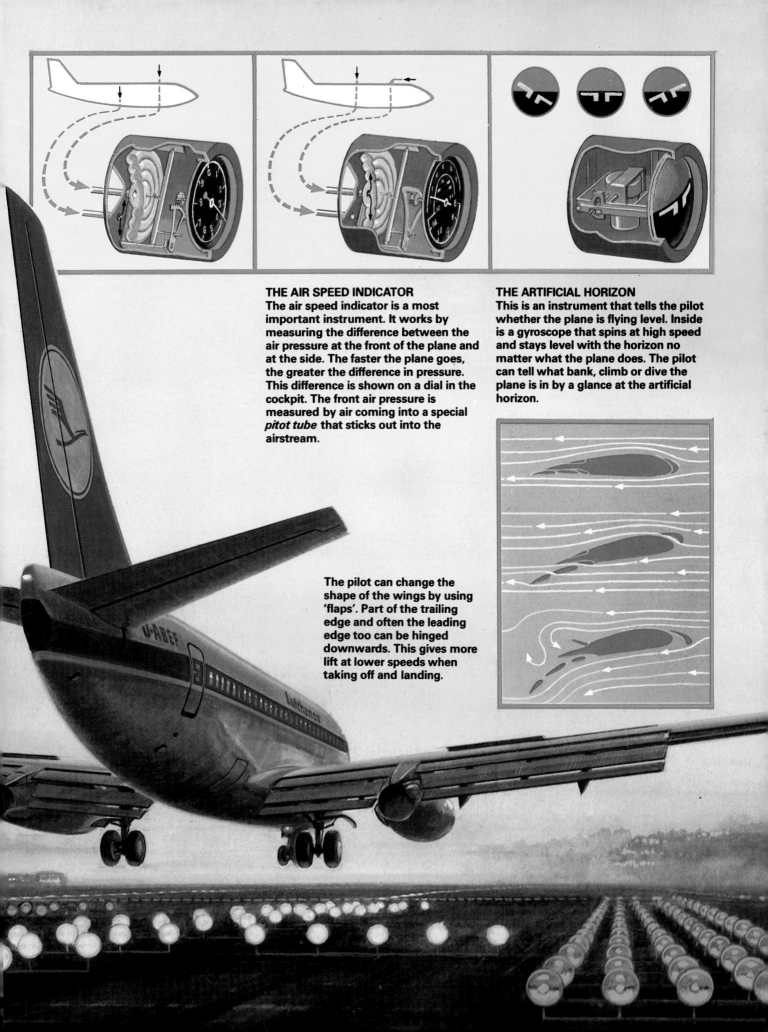

## THE AIR SPEED INDICATOR

The air speed indicator is a most important instrument. It works by measuring the difference between the air pressure at the front of the plane and at the side. The faster the plane goes, the greater the difference in pressure. This difference is shown on a dial in the cockpit. The front air pressure is measured by air coming into a special *pitot tube* that sticks out into the airstream.

## THE ARTIFICIAL HORIZON

This is an instrument that tells the pilot whether the plane is flying level. Inside is a gyroscope that spins at high speed and stays level with the horizon no matter what the plane does. The pilot can tell what bank, climb or dive the plane is in by a glance at the artificial horizon.

The pilot can change the shape of the wings by using 'flaps'. Part of the trailing edge and often the leading edge too can be hinged downwards. This gives more lift at lower speeds when taking off and landing.

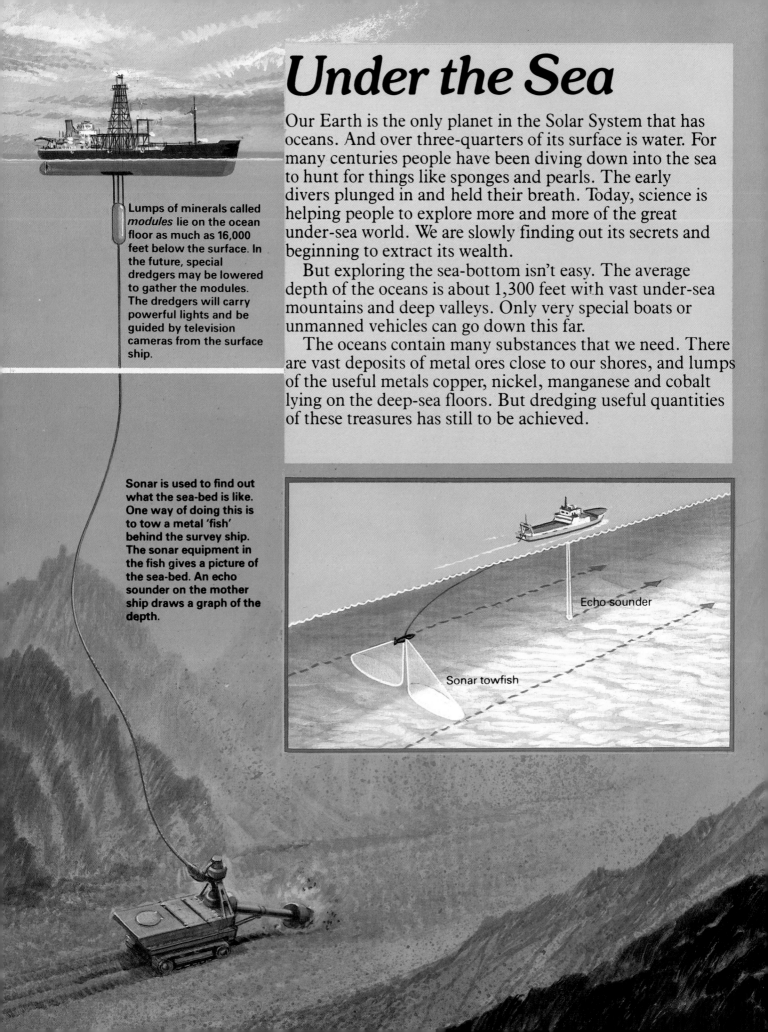

# Under the Sea

Our Earth is the only planet in the Solar System that has oceans. And over three-quarters of its surface is water. For many centuries people have been diving down into the sea to hunt for things like sponges and pearls. The early divers plunged in and held their breath. Today, science is helping people to explore more and more of the great under-sea world. We are slowly finding out its secrets and beginning to extract its wealth.

But exploring the sea-bottom isn't easy. The average depth of the oceans is about 1,300 feet with vast under-sea mountains and deep valleys. Only very special boats or unmanned vehicles can go down this far.

The oceans contain many substances that we need. There are vast deposits of metal ores close to our shores, and lumps of the useful metals copper, nickel, manganese and cobalt lying on the deep-sea floors. But dredging useful quantities of these treasures has still to be achieved.

Lumps of minerals called *modules* lie on the ocean floor as much as 16,000 feet below the surface. In the future, special dredgers may be lowered to gather the modules. The dredgers will carry powerful lights and be guided by television cameras from the surface ship.

Sonar is used to find out what the sea-bed is like. One way of doing this is to tow a metal 'fish' behind the survey ship. The sonar equipment in the fish gives a picture of the sea-bed. An echo sounder on the mother ship draws a graph of the depth.

Echo sounder

Sonar towfish

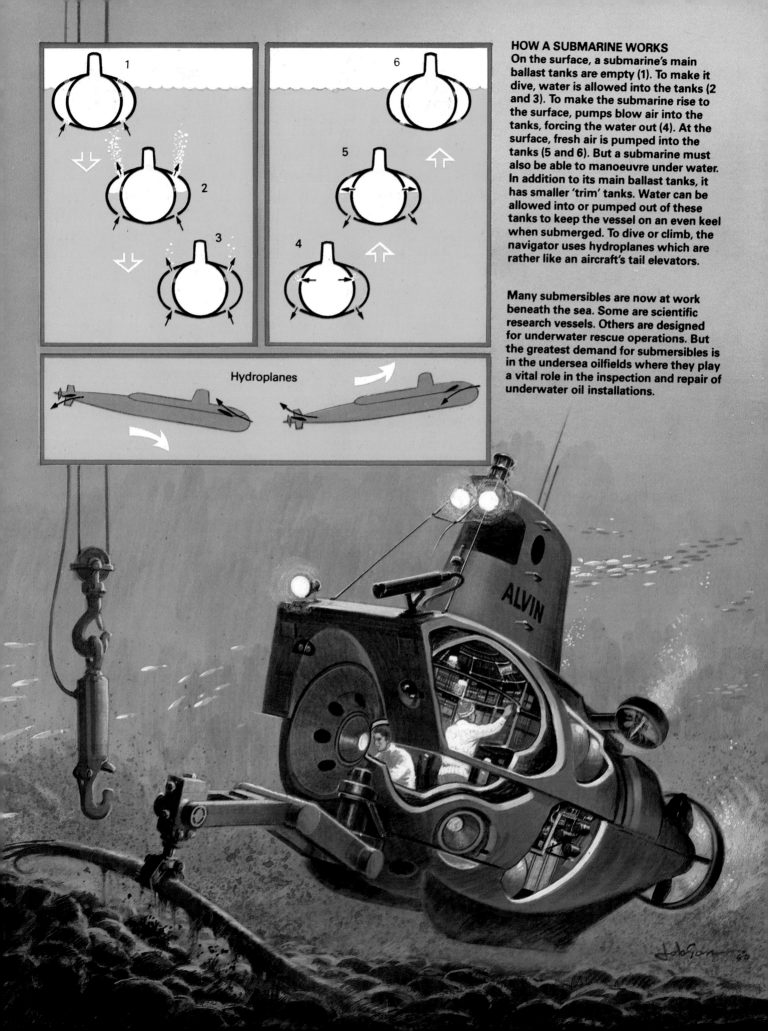

## HOW A SUBMARINE WORKS

On the surface, a submarine's main ballast tanks are empty (1). To make it dive, water is allowed into the tanks (2 and 3). To make the submarine rise to the surface, pumps blow air into the tanks, forcing the water out (4). At the surface, fresh air is pumped into the tanks (5 and 6). But a submarine must also be able to manoeuvre under water. In addition to its main ballast tanks, it has smaller 'trim' tanks. Water can be allowed into or pumped out of these tanks to keep the vessel on an even keel when submerged. To dive or climb, the navigator uses hydroplanes which are rather like an aircraft's tail elevators.

Many submersibles are now at work beneath the sea. Some are scientific research vessels. Others are designed for underwater rescue operations. But the greatest demand for submersibles is in the undersea oilfields where they play a vital role in the inspection and repair of underwater oil installations.

Hydroplanes

1  2  3  4  5  6

ALVIN

# The World of Speed

About a hundred and fifty years ago, the fastest way to travel was on a horse. It had been like this for thousands of years until the invention of the steam engine. Stephenson's *Rocket* locomotive of 1829 had a top speed of 28 mph, just faster than the top speed of a horse-drawn chariot. But during the 19th century, the steam engine was the 'king of speed' throughout the world. By 1850, trains could travel at more than 62 mph and rail travel was available to everyone.

Nowadays, high-speed travel is part of our daily lives. many people are fascinated by speed and enjoy speed sports such as motor racing. Modern electric trains hurtle along special tracks at over 186 mph. The *Concorde* flashes across the Atlantic at twice the speed of sound. Good highways allow us to get from place to place easily and quickly.

The first aircraft were slow and clumsy. Their small engines gave them just enough speed to stay in the air. But as time went by, planes could fly faster and faster, until they reached speeds that were as fast as propeller-driven planes could go — about 1,100 mph. It was the jet engine that made higher speeds possible.

The first plane to fly faster than the speed of sound (about 700 mph) was the rocket-powered Bell X-1, in 1947. Today, planes like the Blackbird SR-71A (above) can fly at speeds of about 2,100 mph. They can also fly at very great heights. This plane has reached a height of over 78,000 feet.

Wind tunnels are used to test the action of air against vehicles. Wind is blown through the tunnel at different speeds so that the scientists can see how the aircraft or train will react. In supersonic tunnels, special instruments show changes in the density of air as it flows around the model (below, left).

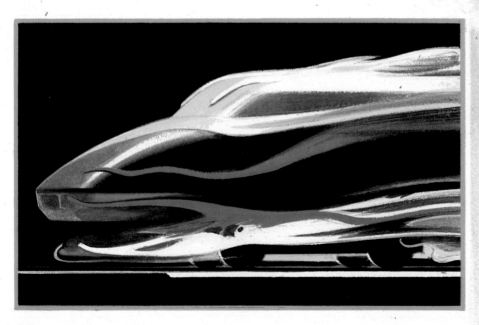

A ship traveling in water cannot go much faster than about 55 mph. But craft that skim on or above the surface can travel much faster. A hydrofoil (right) sits on the water at low speed. But as it goes faster it is lifted until it is skimming along on long legs, rather like a person on skis. This cuts down the 'drag' of the water and hydrofoils can travel at around 70 mph, provided the sea is not too rough. Most modern hydrofoils have 'surface-piercing' foils. These are shaped like a shallow V and they keep the craft stable as it turns sharply or speeds through the waves.

Hovercraft are also skimmers. They travel along on a cushion of air that lifts the craft clear of the water. Big passenger-carrying hovercraft can speed along at more than 85 mph.

The world land speed record has advanced slowly over the years. In 1907 the record was held by a steam-driven car called *Wogglebug*. It reached a speed of 150 mph. Today, the world's record is held by Britain's Richard Noble. In 1983 he reached a speed of 634 mph in his jet-powered *Thrust 2* (below).

By far the fastest travelers are astronauts. Their rocket-powered spaceships have to reach a speed of almost 25,000 mph to get away from the Earth's gravity. We see pictures of Shuttle astonauts out in space, slowly and carefully edging themselves about with their spacepacks. But it is difficult to appreciate that they and their ship are still traveling at 20 times the speed of *Concorde*. Up in space there is no air or anything else to give the astronauts any idea of speed.

But even astronauts are very slow-moving compared to the fastest thing we know about – the speed of light. It travels 186,322 miles in just one second!

# Canals, Bridges and Tunnels

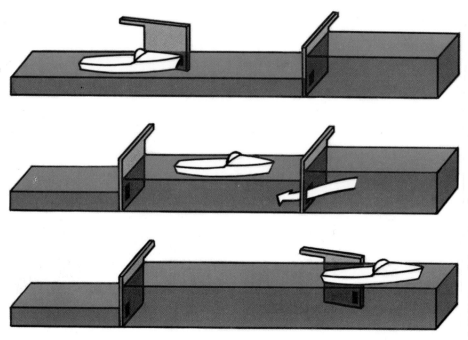

## HOW LOCKS WORK
A lock is like a step. It raises ships and boats in a canal or river from one level to another. If a boat is climbing, it enters the first lock. The lower gate is open and the water inside the lock is at the same level as the lower part of the canal. When the boat is in the lock, the lower gate is closed. A small sluice (a flap) is opened in the upper gate and water flows down. The boat rises as the water level rises. When the water level is the same as the upper level, the upper gate is opened and the boat sails off. Ships can go through a series of locks to raise them to the required level. This happens in big canals such as that at Panama between North and South America. The Panama Canal has three sets of locks, built in pairs so that ships can pass through in both directions at the same time. Each lock is 1,000 feet long and 110 feet wide. Electric locomotives pull ships through the lock system.

Builders of roads and railways often reach mountains or rivers that they have to tunnel through or bridge. People who build bridges and dig tunnels and canals are called civil engineers.

Bridges are among the most spectacular of man-made structures. The simplest and oldest kind of bridge is the *beam bridge*. This is just a wooden or steel beam supported at either end by piers. The length or span of a beam bridge

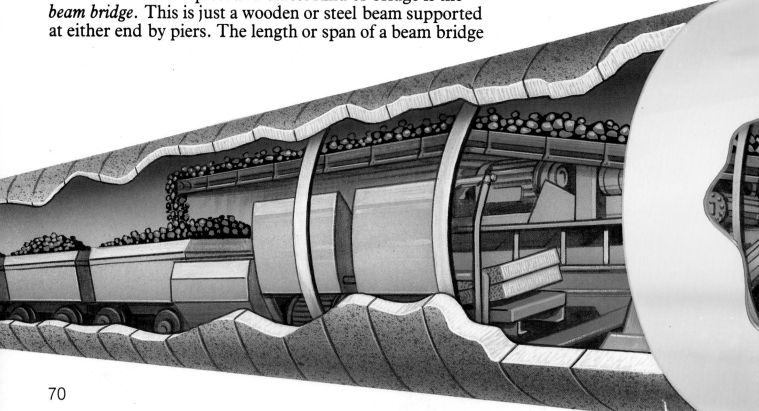

**Beam bridge**          **Arch bridge**          **Suspension bridge**

cannot be too great, otherwise the beam would sag and give way in the middle. Longer beam bridges have several piers along their length.

The *arch bridge* can have a longer span than a beam bridge. This is because the weight of the bridge travels down the sides of the arch to the supports at each end. The ancient Romans were expert builders of arch bridges.

The longest spans of all can be achieved with the *suspension bridge*. In this kind of bridge the road hangs in the air from a pair of heavy cables slung between high towers at either end. The cables are made up of thousands of steel wires bundled together until they make up a cable as much as a yard thick. These cables pass up and over the towers and then down to firm anchorage points on the ground.

**Some Famous Bridges**
The world's longest single bridge span is the main span of the bridge over the Humber estuary in Britain. It is 4,622 feet long and is supported between two huge towers 550 feet high.

The longest suspension bridge, measured between anchorages at either end, is the Mackinac Straits bridge in Michigan, USA. It measures 8,330 feet.

The longest steel arch bridge is the New River Gorge bridge at Fayetteville, West Virginia, USA. It has a span of 1,700 feet.

Probably the most famous bridge in the world is the Golden Gate bridge at San Francisco. Its span is 4,200 feet.

The Verrazano Narrows bridge joins Brooklyn and Staten Island at the entrance to New York harbor. This great bridge has two decks, each with six lanes of traffic.

The world's widest long-span bridge is Australia's Sydney harbor bridge. It is 160 feet wide and has eight lanes of traffic and a footway.

Although we are seldom aware of it, the ground under our feet is often full of tunnels. Tunnels carry electricity cables, gas pipes, water and sewage. Other larger tunnels take underground railways, and tunnels through hills and mountains carry trains and cars. There are different ways of digging tunnels. If the tunnel is to be near the surface, the engineers can dig a deep trench, put a big tube in the trench and cover over the tube again. This method is called 'cut and cover'. The more modern way to dig a tunnel is to use a 'mole' like the one shown here. A powerful cutting head made of steel wheels is pushed and turned forward by powerful electric motors. The cutting head grinds up the earth and rock and this is passed back on conveyor belts to trucks that carry the rubble out of the tunnel. A steel shield protects the men working near the cutting head. As the head moves forward the tunnel is permanently lined with either steel or concrete.

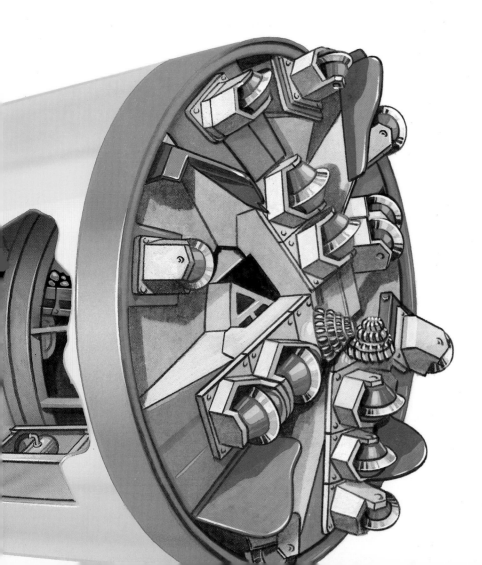

# Fossil Fuels

A fuel is something that can be burned to give heat, light or power. It is a store of energy. The energy came in the first place from the Sun. Plants gather energy from the Sun. Oil, gas and coal were formed from plants that lived millions of years ago. They are called *fossil fuels*.

Drilling for oil is a very costly operation. First of all, the oil has to be found. Geologists spend a long time examining the layers of rock under the land or under the sea-bed. When they think there may be oil, an exploration well is drilled. If oil is found, then several more wells are drilled to find out if there is enough oil to justify a full-scale operation. A production platform is put in place, and only then does the valuable oil start to flow. In the stormy North Sea off the coast of Britain, the platforms may be in 1,000 feet of water. The oil itself can be 10,000 feet below the sea-bed.

Mud tank

As the drill chews into the rock, special 'mud' is pumped down the pipe. The mud comes back up to a tank, where it is cleared of debris. The mud keeps the drill cool and coats the inside of the drilled hole

## How Coal is Formed

Like oil, coal is formed from living things. It started off millions of years ago as trees and plants in ancient forests. The forests slowly sank into swamps and were covered by layers of mud which later became solid rock. The pressure of this rock and the heat from the Earth began to change the plant remains.

The first stage in the formation of coal can be seen in some wet moorlands and bogs. There decaying plants form *peat*, a substance that can be cut and dried to make fuel that burns.

If peat is left in the ground for long enough it becomes *lignite* or brown coal. As more millions of years go by, the coal grows into *bituminous* coal, the black stuff we burn. The last stage in coal formation is *anthracite*, a shiny black rock that is clean to handle. Anthracite is almost pure carbon.

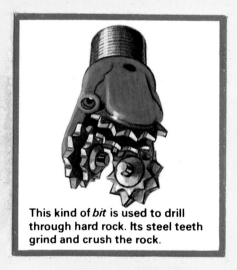

This kind of *bit* is used to drill through hard rock. Its steel teeth grind and crush the rock.

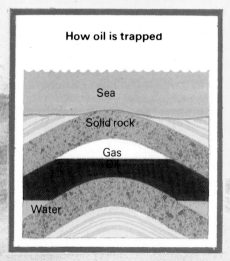

How oil is trapped

Sea

Solid rock

Gas

Oil

Water

## How Oil Forms Underground

The oil we use today was probably formed from decayed plants and animals that fell to the ocean floor 500 million years ago. Slowly, over thousands of centuries, the remains of the plants and animals were covered by layer upon layer of sand and mud. The pressure of these layers caused great heat. This heat, combined with chemical action, changed the ancient remains into the substances we call oil and gas. As more time went by, the oil and gas seeped slowly upwards through soft rock. After a while, they reached solid rock and they could go no further. They were in a *trap*, in which collected oil, gas and water. It is these traps which today's oilmen search for.

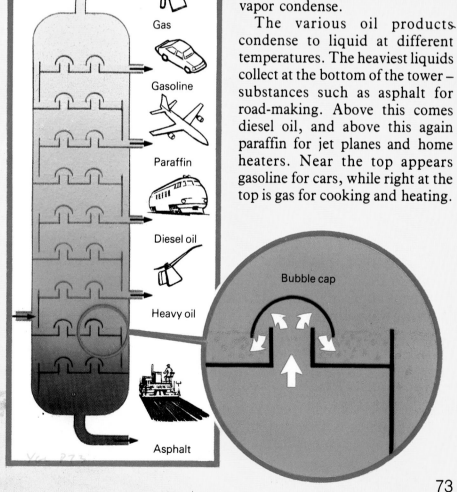

Gas

Gasoline

Paraffin

Diesel oil

Heavy oil

Asphalt

Bubble cap

## All Kinds of Oil

Petroleum from the ground is often called 'crude oil' because it is a complicated mixture of chemicals that has to be sorted out before we can use all the materials in it. This sorting out is done at an oil refinery. A refinery is a maze of steel towers and pipes, but among the main pieces of equipment are tall *fractionating columns*.

Crude oil is heated to a vapor in a furnace. A pipe carries the vapor to an entrance near the bottom of the column. Inside the column, trays are arranged one above the other (see below left). The trays are hottest at the bottom and coolest at the top of the tower. As the oil vapor rises, it passes through holes and is caught by *bubble caps* like upside-down cups (see inset below). The caps force the vapor down again through liquids that have already condensed in the trays. This makes more of the vapor condense.

The various oil products condense to liquid at different temperatures. The heaviest liquids collect at the bottom of the tower – substances such as asphalt for road-making. Above this comes diesel oil, and above this again paraffin for jet planes and home heaters. Near the top appears gasoline for cars, while right at the top is gas for cooking and heating.

73

# Power of the Atom

Scientists have discovered how to get vast amounts of power from the atom. This knowledge can be used for the good of mankind – to make electricity. It can also be used to make nuclear weapons.

The nucleus (center) of the atom is made up of particles called protons and neutrons, held together by powerful forces of attraction. When the nucleus is broken apart, these forces are released as energy in the form of radiation.

The atoms of some elements are constantly breaking up by themselves. They are called *radioactive elements*. Uranium is one of these. One kind of uranium has 92 protons and 143 neutrons in its nucleus. We add these numbers together and call this uranium 235 or U-235 for short. In U-235, a neutron sometimes shoots off from the nucleus. Sometimes this neutron hits another U-235 neutron and knocks loose another neutron. If this should happen often enough we have a *chain reaction* and a great deal of energy is let loose. If a piece of U-235 is large enough, a great many neutrons fly about at once. The reaction gets out of hand and we have a tremendous explosion – an atomic bomb.

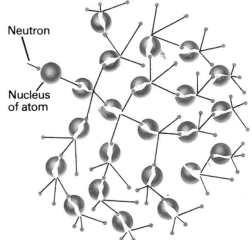

Neutron

Nucleus of atom

**A chain reaction occurs when a neutron splits a uranium atom and produces at least two more neutrons which in turn split other atoms. An uncontrolled chain reaction produces a nuclear explosion.**

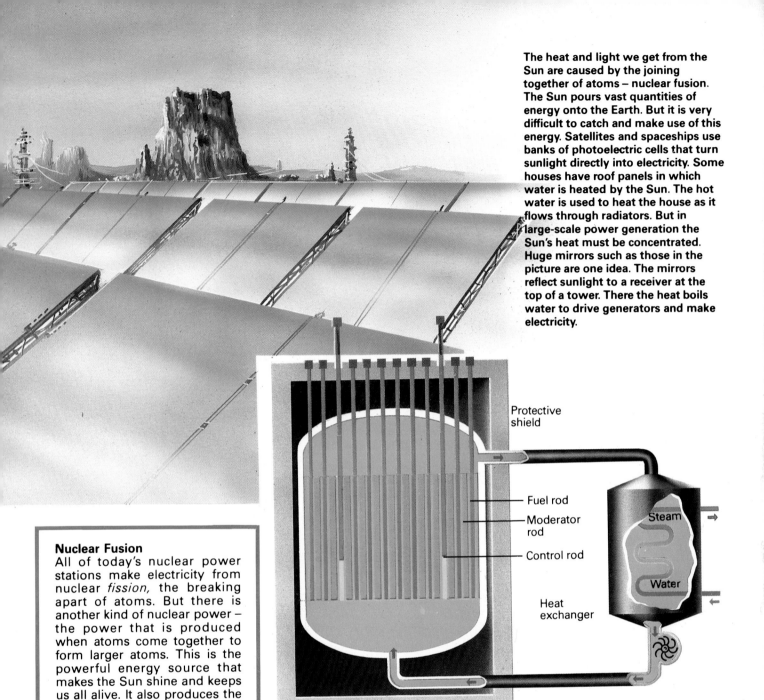

The heat and light we get from the Sun are caused by the joining together of atoms – nuclear fusion. The Sun pours vast quantities of energy onto the Earth. But it is very difficult to catch and make use of this energy. Satellites and spaceships use banks of photoelectric cells that turn sunlight directly into electricity. Some houses have roof panels in which water is heated by the Sun. The hot water is used to heat the house as it flows through radiators. But in large-scale power generation the Sun's heat must be concentrated. Huge mirrors such as those in the picture are one idea. The mirrors reflect sunlight to a receiver at the top of a tower. There the heat boils water to drive generators and make electricity.

Protective shield

Fuel rod

Moderator rod

Control rod

Heat exchanger

Steam

Water

## Nuclear Fusion

All of today's nuclear power stations make electricity from nuclear *fission,* the breaking apart of atoms. But there is another kind of nuclear power – the power that is produced when atoms come together to form larger atoms. This is the powerful energy source that makes the Sun shine and keeps us all alive. It also produces the terrible power of hydrogen bombs. It is called nuclear *fusion.*

In nuclear fission the atoms of heavy elements such as uranium and plutonium are split apart. In nuclear fusion, the atoms of light elements such as hydrogen and helium are forced together. Inside the Sun, it is the turning of hydrogen into helium that produces our light and heat.

Scientists have been trying for many years to make nuclear fusion work safely to produce all the energy we need. If they succeed, all our fuel shortages and power problems will end.

In a nuclear power station, the chain reaction is controlled. A nuclear reactor has rods containing some U-235 placed in a *moderator.* The moderator is usually graphite or water. In case the chain reaction starts to go too quickly, *control rods* are inserted too. These are made of metals that absorb neutrons and can be moved in or out of the reactor.

Heat is produced by the nuclear reaction and a cooling liquid or gas passes through the reactor to take up the heat. This coolant goes to a *heat exchanger* where it makes steam to drive steam turbines. These turbines make electricity.

# Power of the Sea

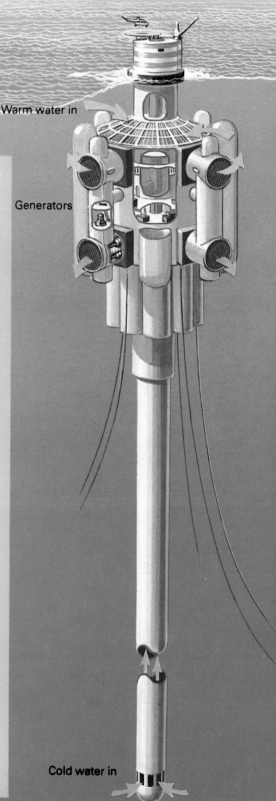

Warm water in

Generators

In a few years from now, the world's oil wells will begin to dry up. We will have to find other kinds of energy, and scientists are turning their attention to some of the Earth's 'free' energy sources. Among those 'free' sources is the energy that can be got from the sea.

If you have been at the seaside when big waves are coming in you will know the power behind these moving masses of water. There can, in fact, be 80 kilowatts of power for every yard of wave along its length. Many different ways of collecting power from the waves have been tried, but so far none of them has been completely successful. People have tried floating 'ducks' that move up and down in the waves. Machines inside the 'ducks' turn the bobbing motion into electricity. Others have tried hinged rafts which move with the waves and drive turbines.

The ocean currents are another possible source of power. Year in, year out, great masses of water move steadily through the oceans, always following the same path. The Gulf Stream that runs across the Atlantic Ocean towards northern Europe is such an ocean current. Scientists have been working on machines to make use of this vast mass of moving water. But the ocean currents move slowly, so the machines will have to be very big to work well (see opposite page).

Another idea, pictured on the right, is a huge floating power station. As the Sun beats down on the oceans, it heats the surface waters, but the water deep down stays very cold. This difference in temperature between the two layers of water can be used to work a heat engine. Cold water goes in at the bottom and warm surface water goes in at the top. The difference in temperature is made to drive turbo-generators to make electricity.

Cold water in

**Above:** Lines of hinged rafts are moored facing the direction from which the wind usually blows. As the waves pass under them, the rafts move up and down, making the hinges work as pumps. The pumps produce liquid at high pressure to drive an electric generator.

At the right above is a floating power station. It is here that the energy from the rafts is turned into electricity before being sent ashore by cable.

**Right:** A possible scheme to get power from the Gulf Stream or other ocean currents. It is a huge underwater turbine, with blades hundreds of feet across. The turbine has to be very big because ocean currents move quite slowly. You can tell the size of the turbine by comparing it with the submersible.

Another wave-power machine is called the *oscillating water column.* This is simply a long vertical tube open to the air at the top, with the bottom below the surface. When a wave passes, the water inside the tube rises and falls. The rise and fall of water is like a piston in a cylinder which can drive an air turbine to make electricity.

**Left:** A floating thermal power station in which differences in water temperature are used to drive electricity generators.

All of these schemes are future possibilities. The way that the energy of the sea has been harnessed already is in tidal power stations where the difference in water level between high and low tides is used to power turbines.

# Measuring Time

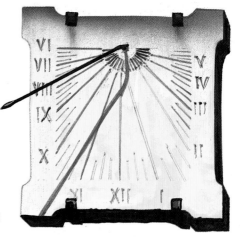

For centuries, people used the Sun, Moon and stars to tell the time by. It takes a year for the Earth to go right round the Sun once. The Moon's monthly phases give us the 12 divisions into which the year is divided. But, more important, the Earth rotates once on its axis every 24 hours, giving regular day and night.

For thousands of years, these rough divisions of time were accurate enough. Then came the sundial, the water clock and the sandglass.

The first mechanical clocks appeared about the end of the 13th century, but they were not very accurate. It was not until the pendulum clock was invented in 1656 that people could really tell the time with any certainty. The pendulum clock is based on something discovered earlier by the great Italian scientist Galileo. When Galileo was a boy, he watched the swinging of a hanging lamp in Pisa cathedral. He found that each swing of the lamp as it moved back and forth took exactly the same time, whether it was a big swing or a small one. Many accurate pendulum clocks are still in use today.

Today's very accurate clocks and watches are not mechanical. Electric watches run off tiny batteries. There are quartz clocks, driven by the tiny vibrations given off by quartz crystals when an electric current is applied to them. Atomic clocks are driven by the movement of atoms and molecules. These can be accurate to less than one thousandth of a second a year.

**Sundials were first used by the Egyptians. As the Sun moves across the sky, its shadow shows the time of day.**

**So that dawn and sunset fall at about the same time on the clock all over the world, there are many time zones. Most of them are one hour apart. Large countries such as the United States have many time zones. When you go from west to east, you have to advance your watch every time you enter a new time zone. If you cross the International Date Line on the 180° meridian your watch goes back 24 hours to make up for the time difference.**

***Concorde* crosses the Atlantic so quickly, it 'beats' the clock.**

**The Egyptians developed the water clock in about 1400 BC. This clock had holes in the bottom that allowed water to drip slowly out. Levels were marked inside to show the time.**

The early mechanical clocks were powered by weights. The weights were attached to a cord that was wound round a drum. As one weight fell, the drum turned. This worked a set of toothed wheels called gears. One of these gears turned the clock hands. Another toothed wheel drove a lever called an escapement. The escapement rocked back and forth and controlled the speed of the gears.

Clocks driven by springs appeared in the 15th century. These were more accurate than the weight-driven timepieces.

**The sand-glass had two containers joined by a narrow stem. It took a fixed length of time for the sand to trickle from one container to the other.**

Very accurate time-keeping was needed at sea. The first accurate chronometers were made by John Harrison. The one above was regulated by a balance spring.

The modern digital watch shows the time in numbers. At the heart of the watch is a quartz crystal. An electric current from a tiny battery is fed to the crystal. This makes it vibrate at a steady frequency. The crystal gives out electric signals at this frequency which go to the microchip part of the watch. The microchip counts the signals and every second, minute and hour sends another signal to the digital display. It may also display the date.

The inside of an electronic watch.

A clock made in 1580.

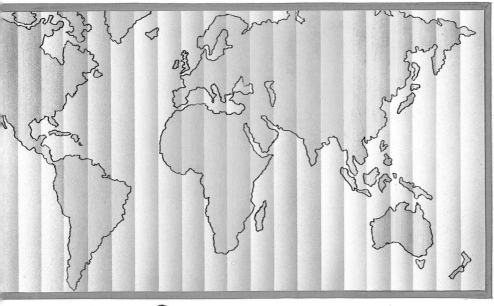

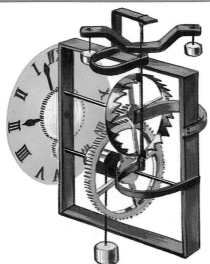

The early clocks were driven by a falling weight and regulated by a rocking bar. The bar allowed a wheel connected to the hands to move round one tooth at a time.

Pendulum clocks have an escapement like the one above. A pair of teeth attached to the pendulum allow a wheel to turn one tooth at a time.

# Science in the Home

Science has played a big part in making our homes comfortable and easy to operate. Even forgetting such things as television sets, radios, refrigerators and food mixers, there are many other ways in which science works for us. Many of the home inventions of the past hundred years are now so much a part of our daily lives that we seldom stop to think about them. The first canned food, for example, went on sale in the 1820s. But it was not until the 1860s that the first can-opener was invented. Before then, people opened cans with a hammer and chisel!

Cleaning and washing used to be hard work, often needing servants armed with mops and brooms. Then science came to our aid. The carpet sweeper was invented in

It is very seldom we have to think about all the wires, pipes and tanks that are hidden behind the walls and under the floors of our homes. Some of them are shown in the picture below. The big tank feeds all the water pipes throughout the house. The smaller tanks are fed from it and are for the hot water system. The hot water boiler on the ground floor may use gas, coal or oil to heat the water. The electricity supply comes in through a main fuse box. You can see how the wires run from a fuse through all the electric sockets in a room back to the fuse box.

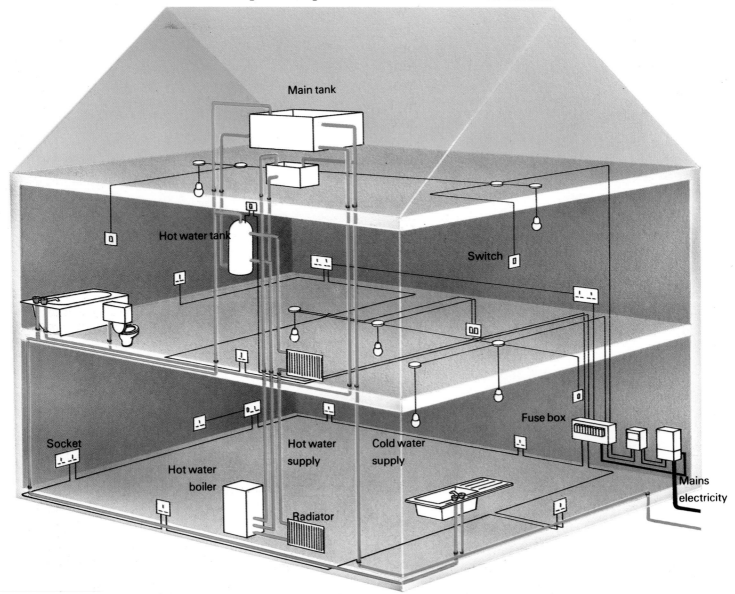

Main tank

Hot water tank

Switch

Socket

Hot water supply

Cold water supply

Fuse box

Hot water boiler

Radiator

Mains electricity

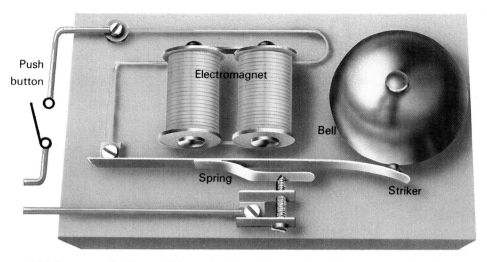

Push button

Electromagnet

Bell

Spring

Striker

## THE ELECTRIC BELL

When you press the button of an electric bell you close a switch that makes an electric current flow through an electromagnet. This magnetism moves a striker which hits the bell. As the striker moves, it opens the contact that passes current to the electromagnet. There is no magnetism, so a spring pulls the striker back. This closes the contact again and current passes again to the electromagnet. The bell rings and opens the contact again. In this way, the bell keeps on ringing as long as the button is pressed.

1876, soon followed by the hand-cranked vacuum cleaner, worked by bellows. In 1901 an electric motor and an air filter turned it into the vacuum cleaner.

The first washing machines were also worked by turning a handle. By 1914, electric motors were being used. Although detergents were invented as long ago as 1916, it was not until 1945 that they came into general use.

The oil lamp, like the candle, was one of the earliest and most useful inventions. Then came gas in the early 1800s. But gas light was rather poor until the invention of the incandescent mantle in 1885. The mantle was a sleeve of fine cotton soaked in chemicals. This fitted over the gas flame and burned to increase and spread the light. The carbon-filament electric lamp, invented around 1878, was the forerunner of today's tungsten-filament lamp and the florescent tube. Many other inventions, big and small, help to make our lives easier; things like safety pins, zip fasteners, sewing machines, safety razors, water closets and non-stick frying pans.

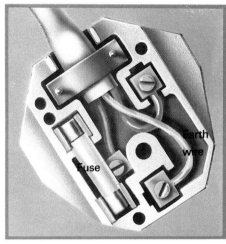

Fuse

Earth wire

Electric plugs (above) should have a fuse inside them for safety. Inside the tube is a thin wire which melts if too much current passes through it. This cuts off the electricity before any damage is done. Without a fuse, an electrical fault could cause the current to rise and cause overheating and start a fire. Fuses are graded in amperes (amps) – an amp is the unit of electric current. The number of amps printed on a fuse is the amount of current that the fuse will allow through to whatever appliance is being used without 'blowing'.

A refrigerator has pipes inside it that contain a cold fluid. This fluid easily changes from a liquid to a vapor. As it goes into the refrigerator it is liquid and is pumped through an evaporator. This lowers the liquid's pressure and it becomes vapor. The change from liquid to vapor makes the vapor cold. This cold vapor flows through pipes inside the refrigerator. After it leaves, it goes to a condenser. This increases its pressure and it changes back to liquid, giving out heat. In this way, heat is taken from inside the refrigerator to the outside.

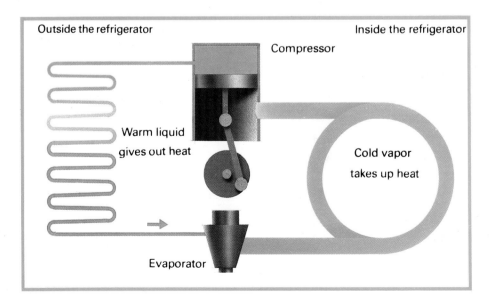

Outside the refrigerator

Inside the refrigerator

Compressor

Warm liquid gives out heat

Cold vapor takes up heat

Evaporator

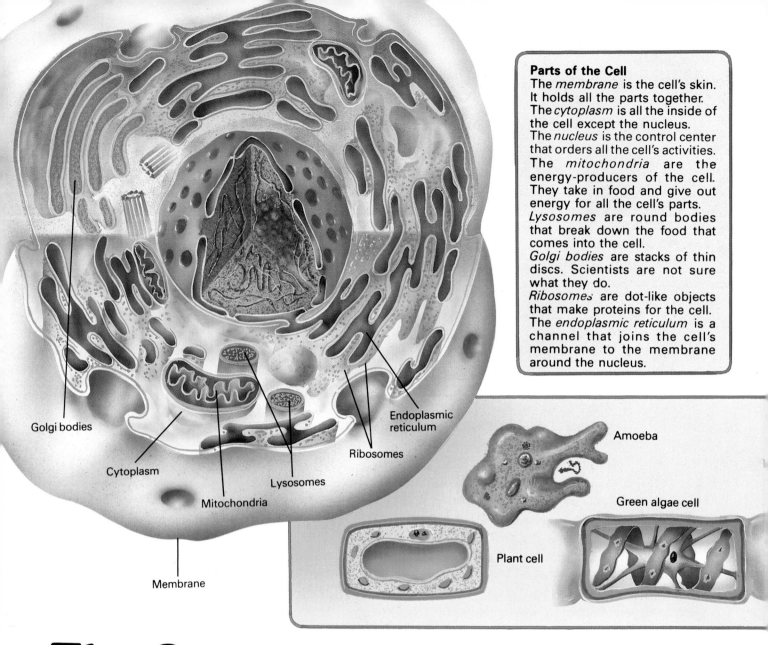

**Parts of the Cell**
The *membrane* is the cell's skin. It holds all the parts together.
The *cytoplasm* is all the inside of the cell except the nucleus.
The *nucleus* is the control center that orders all the cell's activities.
The *mitochondria* are the energy-producers of the cell. They take in food and give out energy for all the cell's parts.
*Lysosomes* are round bodies that break down the food that comes into the cell.
*Golgi bodies* are stacks of thin discs. Scientists are not sure what they do.
*Ribosomes* are dot-like objects that make proteins for the cell.
The *endoplasmic reticulum* is a channel that joins the cell's membrane to the membrane around the nucleus.

Golgi bodies

Cytoplasm

Mitochondria

Lysosomes

Ribosomes

Endoplasmic
reticulum

Membrane

Amoeba

Green algae cell

Plant cell

# *The Science of Life*

Our bodies are made up of millions upon millions of tiny cells. All these cells grow from only two cells that join together at conception, when a new life begins. The cells come in all shapes and sizes. Some brain cells are only 1/5,000 inch across. Other cells, such as muscle and nerve cells, are long and thin. Nerve cells can be as much as a yard long.

But all cells are similar in some ways. They are all enclosed in a thin membrane. Inside the membrane is the *cytoplasm*. This contains several different structures, and each has its own job to do. At the cell's center is its *nucleus*. The nucleus is the cell's 'brain', controlling everything that the cell does.

The nucleus is the cell's control room. Fine strands called *chromosomes* are scattered inside it. Chromosomes are made up of two kinds of substances – DNA and *proteins*. DNA is a chemical that controls all the things that are passed on from one generation to the next. It rules whether we will have fair or dark hair, blue eyes or brown. DNA does this by controlling the production of other substances called proteins.

# Cells at Work

Below you can see some of the different kinds of living cells. *Red blood cells* are those that make our blood look red. They carry oxygen from the lungs all over our bodies. *White blood cells* are those that we need to protect us from disease. *Nerve cells* are shaped as they are so that they can carry messages to different parts of the body. *Lymphocytes* and *phagocytes* are two kinds of cell which 'eat' harmful bacteria in the body. When we cut ourself, bacteria often attack the wound. Then phagocytes come along and attack and surround the bacteria. If the invading bacteria or micro-organisms are still not beaten, the lymphocytes come along and secrete *antibodies* that make the invaders harmless.

Amoebas are tiny creatures that consist of only one cell. All the machinery for life is held inside them. Despite their small size, they are able to move about, feed themselves and reproduce.

## Passing On The Information

A new life begins when two cells join to make a new one. This new cell must have in it all the characteristics that a person inherits from his or her parents. The cell must also contain all the information needed to build up a human body.

Each chromosome in the cell's nucleus is a chain of things called *genes*. There may be 1,000 of them on each chromosome. The genes govern our hair color, eye color and all our other characteristics. The genes are made of DNA. Each DNA molecule is arranged in two spirals linked by chemicals, rather like the rungs of a ladder (see below). There are only four different chemical rungs but they can be arranged in a huge variety of combinations to give all the different characteristics that people have.

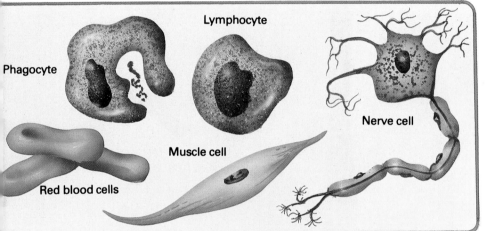

Phagocyte
Lymphocyte
Nerve cell
Muscle cell
Red blood cells

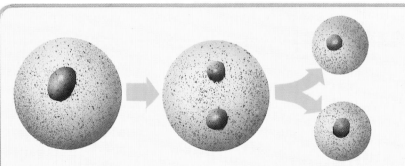

## Making New Cells

The only way new tissue can be formed is from the division of cells. The nucleus splits in half to make two identical nuclei, each with a full set of chromosomes. Then the cell itself divides to make two cells.

There is another kind of cell division which takes place in *sex* cells. In this process, the parent sex cell, male or female, divides in such a way that each new cell has only half the number of chromosomes of the parent cell. So when the sex cells meet, each gives a half set of chromosomes to the new cell, which must therefore have a complete set of chromosomes, half from the father and half from the mother.

# The Body Machine

The human body is a wonderful machine. It is made up of millions upon millions of tiny cells which group together to form organs such as the heart and lungs. Inside a thin covering of skin and a framework of bones are all the different parts that allow us to move, breathe, speak and eat. Our eyes, ears and other senses tell the brain what is going on outside our bodies. The brain controls everything that we do.

We breathe in air so that our bodies can have a constant supply of oxygen. Without this gas, all our cells would quickly die. We eat to give our bodies fuel.

Our lungs are like a pair of bellows in our chest. We have a sheet of muscle called the *diaphragm* stretched across the bottom of our chest. When this and the rib muscles contract, the space in our chest grows bigger. Air rushes in through our nose and into our lungs. When air goes into our lungs, it passes through finer and finer tubes until it enters tiny air sacs called *alveoli*.

The alveoli are covered with a network of fine blood vessels. Oxygen passes from the air into the blood and the blood gives back carbon dioxide gas to the lungs. The blood carries oxygen to all parts of our bodies.

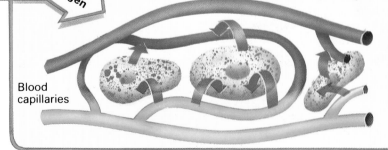

Air in

Carbon dioxide out

Blood with oxygen

Alveoli

Blood without oxygen

Lung

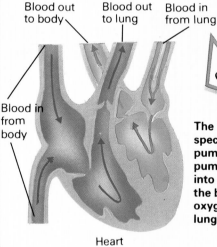

Blood out to body

Blood out to lung

Blood in from lung

Blood in from body

Heart

Oxygen

The heart is a pump, but a very special one. It has two separate pumping systems. The left side pumps oxygen-carrying blood into our *arteries* and all around the body. The right side pumps oxygen-free blood back to the lungs.

Food

Blood is red because it contains millions of tiny discs which we call the red blood cells. It is these cells that carry the oxygen around our bodies. Blood also contains white cells that help to defend the body against disease. The red and white cells float in a watery liquid called *plasma*.

Oxygen

Blood travels all over the body through a vast network of blood vessels. First of all it goes through our arteries. Then it travels through finer and finer tubes until it reaches very fine vessels called *capillaries*. The capillaries give food and oxygen from the blood to all the body's cells. In return, the cells give the capillaries carbon dioxide and other waste materials that travel back to the heart through our *veins*.

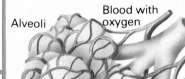

Blood capillaries

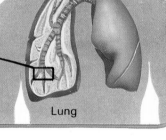

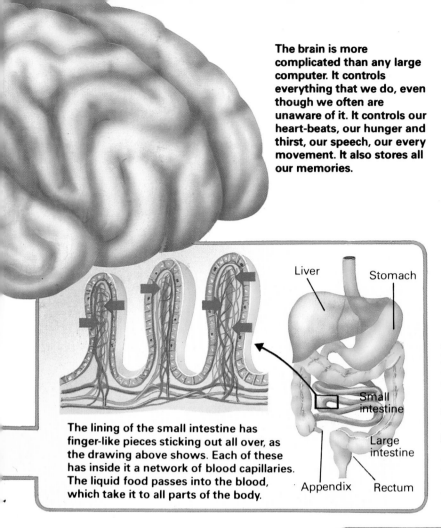

The brain is more complicated than any large computer. It controls everything that we do, even though we often are unaware of it. It controls our heart-beats, our hunger and thirst, our speech, our every movement. It also stores all our memories.

Liver

Stomach

The lining of the small intestine has finger-like pieces sticking out all over, as the drawing above shows. Each of these has inside it a network of blood capillaries. The liquid food passes into the blood, which take it to all parts of the body.

Small intestine

Large intestine

Appendix    Rectum

## The Nervous System

The nervous system controls all the other systems in our body. It has two main parts. The first part is made up of the brain and the spinal cord that runs down our back inside the backbone. The other part of the nervous system is made up of nerves that go out from the spinal cord and the brain to the various parts of the body.

The nerves are rather like telephone wires. When the doorbell rings, a message goes from our ears to our brain. The brain decides what we should do about it and sends messages to various parts of our body, telling them to work together and get out of our chair. Our muscles contract in the right order; we go to the door and open it.

## Reflex Actions

But not all our actions have to be ordered by the brain. If you touch a very hot plate, you pull your hand away instantly without having to think about it. This is called a *reflex action*. Nerves in our hand detect the heat and shoot a message to a nerve center in the spinal cord. The nerve center immediately sends back a message to the hand to draw itself away.

Our food stays in the *stomach* for some time. There it is squeezed and churned and mixed with juice. Then it passes slowly into the *small intestine*, a coiled tube about 20 feet long. Two glands, the *liver* and the *pancreas*, are connected to the small intestine. The liver pours a liquid called *bile* into the intestine. This helps to digest fat. The pancreas sends in substances called *enzymes* that also help to break down the food.

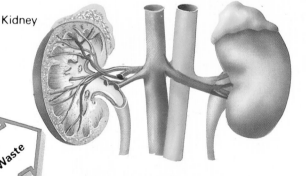

Kidney

Waste

The removal of waste from our bodies is called *excretion*. Several important organs work to get rid of waste. We lose water as sweat through our skin. We breathe out carbon dioxide and water from our lungs.

ood
nd
xygen

Energy

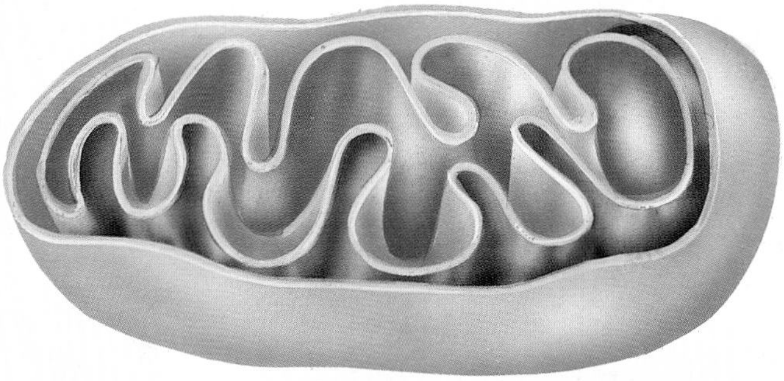

The cells in our bodies have inside them hundreds of tiny sausage-shaped objects called *mitochondria*. (The one in the picture above is very much enlarged.) Mitochondria are power stations. They burn up the food we eat to give us energy. There may be as many as 800 of them in one tiny cell.

Our two *kidneys* (pictured above) get rid of most of our unwanted water and other waste. They filter our blood and we get rid of the waste as urine. The *liver* is the largest gland in our body. As our blood passes through the liver, its poisons and waste are removed. The liver also stores substances that our body uses when they are needed.

# Animals in Motion

If we had to think about all the things that happen to our bodies when we run, we would never be able to do it! Think about just a few of them. We bend and unbend joints in our feet, ankles, knees, hips, shoulders, elbows and wrists. Muscles pull in the right order to make all these joints work at just the right moment. Other muscles make us breathe more deeply. In fact, when we run we use more than a hundred different muscles.

## The Skeleton

Our skeleton is built up of more than 200 living bones. These bones support our body, protect the organs inside us, but allow us to move about freely. Bones meet at movable joints and are held together by bands of tough tissue called *ligaments*. To let the bones move smoothly against each other, their ends are covered with a pad of tissue called *cartilage* (gristle). And the joint is kept 'oiled' by a special liquid.

## The Muscles

All our movements depend on muscles. Muscles work our arms and legs, make us smile, turn our eyes and chew our food. They also pump the blood around our bodies and churn the food in our digestive system.

Muscles are made up of thousands of long, thin fibers. The fibers are arranged in bundles enclosed in a sheath. These muscle fibers contract when they get a signal from a nerve. Muscles work in pairs; one pulls one way and the other pulls the opposite way.

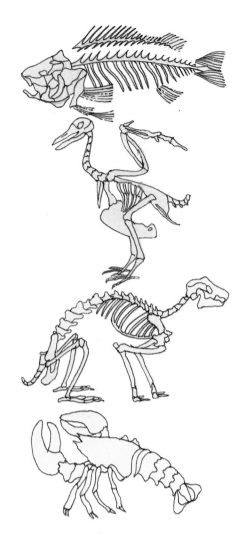

The skeletons of all animals are suited to their way of life. The fish has a 'bendy' backbone. The bird has light bones for flying. The dog has limbs that can move forward and backward, bend and straighten and twist. The lobster has a hard outside skeleton.

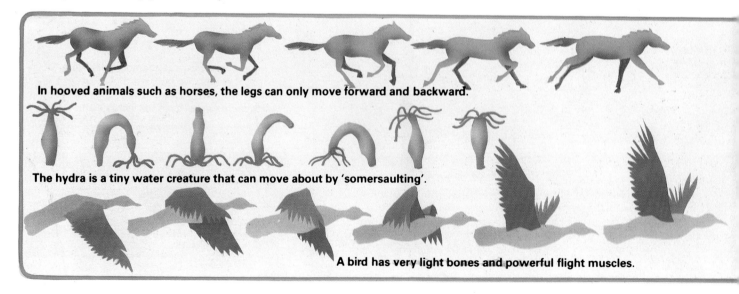

In hooved animals such as horses, the legs can only move forward and backward.

The hydra is a tiny water creature that can move about by 'somersaulting'.

A bird has very light bones and powerful flight muscles.

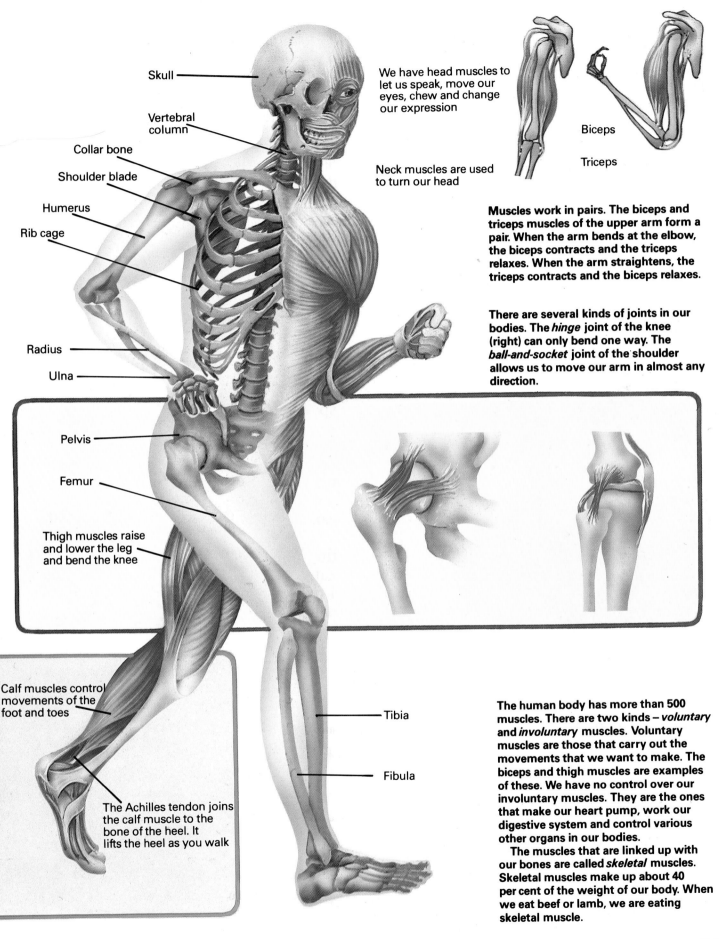

Skull

Vertebral column

Collar bone

Shoulder blade

Humerus

Rib cage

Radius

Ulna

Pelvis

Femur

Thigh muscles raise and lower the leg and bend the knee

Calf muscles control movements of the foot and toes

The Achilles tendon joins the calf muscle to the bone of the heel. It lifts the heel as you walk

We have head muscles to let us speak, move our eyes, chew and change our expression

Neck muscles are used to turn our head

Biceps

Triceps

Muscles work in pairs. The biceps and triceps muscles of the upper arm form a pair. When the arm bends at the elbow, the biceps contracts and the triceps relaxes. When the arm straightens, the triceps contracts and the biceps relaxes.

There are several kinds of joints in our bodies. The *hinge* joint of the knee (right) can only bend one way. The *ball-and-socket* joint of the shoulder allows us to move our arm in almost any direction.

Tibia

Fibula

The human body has more than 500 muscles. There are two kinds – *voluntary* and *involuntary* muscles. Voluntary muscles are those that carry out the movements that we want to make. The biceps and thigh muscles are examples of these. We have no control over our involuntary muscles. They are the ones that make our heart pump, work our digestive system and control various other organs in our bodies.

The muscles that are linked up with our bones are called *skeletal* muscles. Skeletal muscles make up about 40 per cent of the weight of our body. When we eat beef or lamb, we are eating skeletal muscle.

# How Plants Live

All living things are divided into two great groups – the Animal Kingdom and the Plant Kingdom. Animals and plants have several things in common – they grow, they are able to reproduce themselves and make new animals or plants, and they are all made up of tiny cells.

The main difference between animals and plants is in the way they feed. Animals take in ready-made food in the form of plants or other animals. Plants make their own food from simple substances in the air and water. This is called *photosynthesis*. The plants need light for making food. They also need a green substance called *chlorophyll*. This is why most plants are green.

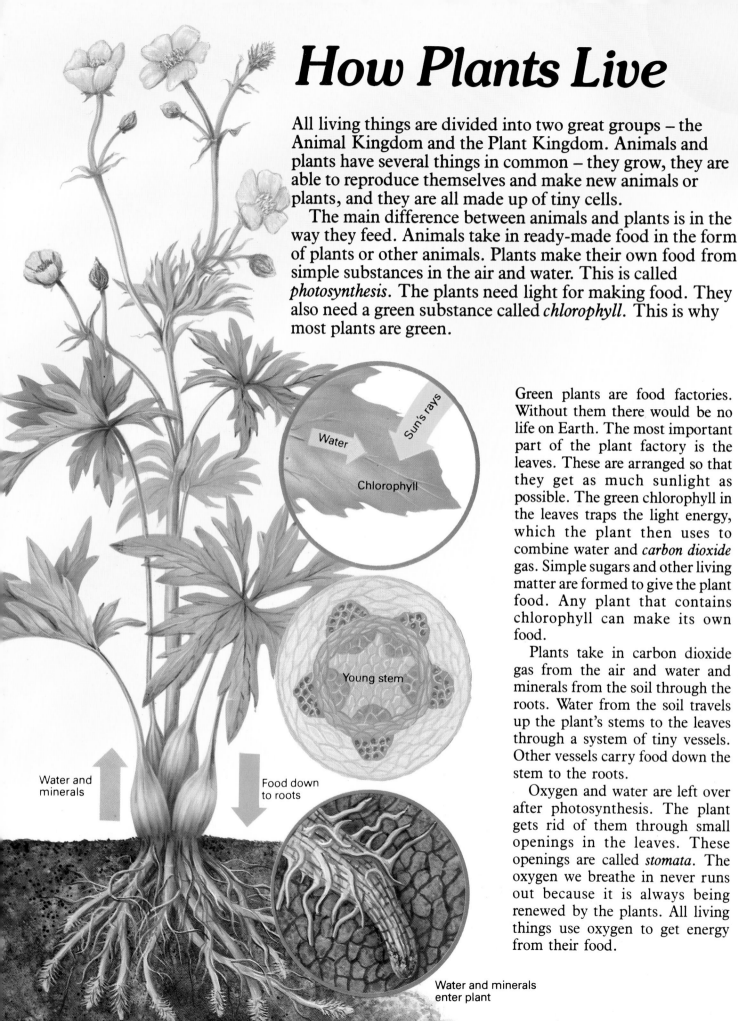

Water

Sun's rays

Chlorophyll

Young stem

Water and minerals

Food down to roots

Water and minerals enter plant

Green plants are food factories. Without them there would be no life on Earth. The most important part of the plant factory is the leaves. These are arranged so that they get as much sunlight as possible. The green chlorophyll in the leaves traps the light energy, which the plant then uses to combine water and *carbon dioxide* gas. Simple sugars and other living matter are formed to give the plant food. Any plant that contains chlorophyll can make its own food.

Plants take in carbon dioxide gas from the air and water and minerals from the soil through the roots. Water from the soil travels up the plant's stems to the leaves through a system of tiny vessels. Other vessels carry food down the stem to the roots.

Oxygen and water are left over after photosynthesis. The plant gets rid of them through small openings in the leaves. These openings are called *stomata*. The oxygen we breathe in never runs out because it is always being renewed by the plants. All living things use oxygen to get energy from their food.

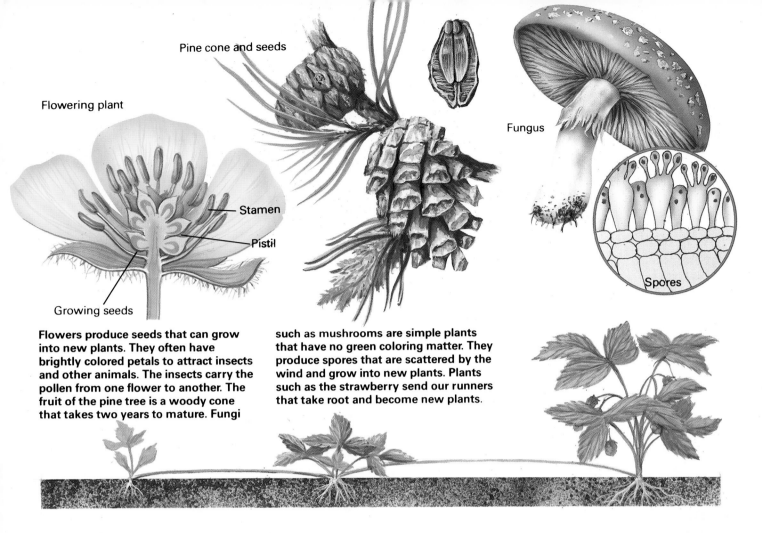

Flowering plant

Pine cone and seeds

Fungus

Stamen

Pistil

Growing seeds

Spores

Flowers produce seeds that can grow into new plants. They often have brightly colored petals to attract insects and other animals. The insects carry the pollen from one flower to another. The fruit of the pine tree is a woody cone that takes two years to mature. Fungi such as mushrooms are simple plants that have no green coloring matter. They produce spores that are scattered by the wind and grow into new plants. Plants such as the strawberry send our runners that take root and become new plants.

## New Plants from Old

Most plants have flowers. The flowers are there to make new plants. Some plants have male and female parts in separate flowers. Other plants have the male and female parts in the same flower. The male parts are called the *stamens*. Male cells – tiny yellow grains called *pollen* – are made in a part of the stamen called the *anther*. The female part of the flower is the *pistil*. It is here that the *ovules* or female egg cells are made.

In *pollination*, pollen grains must be carried from an anther to a pistil. Most plants, including all flowering and cone-bearing plants, reproduce in this way. In some plants, the pollen goes from the anthers to the pistil of the same flower or another flower in the same plant. But in most plants the pollen grains go from one flower's anthers to the pistil of another flower of the same species.

Sometimes the wind carries the pollen from one plant to another. Sometimes an insect such as a bee gets pollen stuck to its body and carries it to another flower.

After pollination, the seeds grow in the flowers and become surrounded by the fruit. The fruit helps to scatter the seeds far and wide so that they have room to grow.

The leaves of some plants have grown in such a way that they can catch insects. These plants live in places where they cannot get enough food from the soil. Insects are attracted by the nectar of the pitcher plant below, but slip down the smooth sides of the pitcher into a pool of liquid which dissolves them.

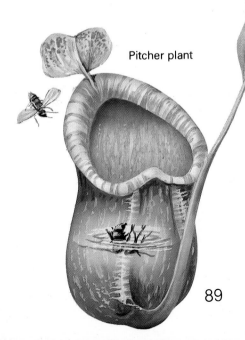

Pitcher plant

89

# Glossary of Science

**Absolute zero**  The lowest temperature that there can be in theory. It is −460°F. Scientists have got to within a few millionths of a degree of this temperature, but will probably never reach it.

**Accumulator**  A device for storing electricity. A common type, the lead-acid accumulator, consists of a container filled with dilute sulphuric acid, in which are lead plates. When a small current is passed through terminals attached to the plates, chemical changes take place in the plates, and the accumulator is 'charged'.

**Acid**  A chemical substance that has a sour taste. Some acids such as sulphuric acid are dangerous, others are harmless, such as the citric acid in oranges and lemons. All acids turn a special sort of paper called litmus paper from blue to red.

**Alkali**  The opposite of an acid. When acids and alkalis are mixed they make a neutral solution to give salt and water. Alkalis turn litmus paper from red to blue.

**Alloy**  A metal made up of more than one element. A dentist's amalgam, for example, is an alloy of 70 per cent mercury and 30 per cent copper.

**Alternating current**  Electric current which rapidly reverses its direction. Mains electricity is usually alternating current (AC) at 50 cycles per second.

**Amplifier**  An electronic device that increases the strength of a signal.

**Archimedes' Principle**  When a body is immersed or partly immersed in a liquid, the liquid buoys up the body with a force that equals the weight of liquid displaced by the body.

**Boiling point**  The temperature at which a liquid changes into a gas.

**Capacitor**  A device for storing charge. It has two metal plates separated by some kind of insulator. It will not allow a direct current to pass, but will allow an alternating current to pass, especially if it alternates very rapidly.

**Catalyst**  A substance that is able to alter the speed of a chemical reaction without itself being changed.

**Combustion**  The chemical reaction in which a substance joins with oxygen and gives off heat and light and burns with a flame.

**Compound**  A substance made up of two or more elements that are combined chemically.

**Conductor**  A substance that allows a free flow of electricity through it.

**Cosmic Rays**  Radiation, mainly in the form of charged particles, that strikes the Earth from outer space.

**Crystal**  A solid substance that has a definite geometrical shape. Most pure solids exist as crystals.

**Diffraction**  The spreading out of light as it passes through a narrow slit or past the edge of an obstacle.

**Electric field**  The region around an electrically charged body.

**Electrolysis**  The conduction of electricity between two electrodes through a solution (electrolyte).

Chemical changes take place at the electrodes.

**Electromagnet**  A magnet with an iron core surrounded by a coil of wire that carries an electric current. The core is only a magnet when the current is switched on.

**Electron**  Negatively charged particle that circles the nucleus in all atoms. In every neutral atom there are as many electrons as there are positive protons in the nucleus.

**Element**  A substance that is made up of exactly similar atoms. An element cannot be split up into simpler substances.

**Energy**  The ability to do work. Energy exists in many forms: heat energy, electrical energy, mechanical energy, chemical energy, etc.

**Fission**  The splitting of the nucleus of an atom which releases vast amounts of energy. Fission takes place in atomic bombs and nuclear reactors. Usually uranium atoms are split.

**Friction**  The force that opposes any attempt to move one surface over another touching it.

**Fusion, nuclear**  The combining of light atoms such as hydrogen to give out vast amounts of energy. Fusion happens in hydrogen bombs and in the stars.

**Gravity**  The force that pulls all masses towards every other mass. Gravity keeps us on the surface of the Earth and the planets circling the Sun.

**Infra-red rays**  Invisible radiations with wavelengths just longer than those of visible light. These are heat rays.

**Insulator** A material that is a poor conductor of electricity, heat or sound.

**Ion** An electrically charged atom or group of atoms.

**Laser** An intense beam of light with all its waves in step.

**Lens** A shaped piece of glass or other transparent material which makes things look larger or smaller by gathering together rays of light or spreading them apart.

**Mass** The mass of an object is the amount of matter in it. We measure mass in pounds and ounces. (See also *weight*.)

**Microwaves** Radio waves with wavelengths less than about 8 inches.

**Molecule** The smallest amount of a substance that can exist alone. It is made up of one or more atoms.

**Neutron** An atomic particle which has no charge. Neutrons are found in the nuclei of all atoms except hydrogen.

**Nucleus, atomic** The center of an atom. It contains protons (positively charged particles) and neutrons (uncharged particles). The nucleus itself is positively charged. Nearly all the weight of an atom is in its nucleus.

**Photon** A quantum or packet of light.

**Plastics** Solid materials that at some stage of their manufacture are plastic (pliable) by heat and pressure. They can then be given a new shape.

**Prism** A triangular block of glass that can split up white light into all the colors of the rainbow.

**Proton** A positively charged particle that is found in the nucleus of every atom. The positive charge of a proton is equal to the negative charge of an electron.

**Radioactivity** A substance whose atoms are always breaking up is said to be radioactive. A radioactive substance usually gives out alpha-particles and beta-particles.

**Reaction** A chemical process that involves two or more substances and which results in a chemical change.

**Reflection** The bouncing back of light rays, sound waves or any kind of electromagnetic radiation after it strikes a surface.

**Refraction** The bending of a light ray as it crosses the boundary between two substances of different optical density.

**Resistance** The property of an electrical conductor that makes it oppose the flow of current through it. Resistance is measured in ohms.

**Salt** A chemical compound that is formed with water when a base reacts with an acid. A salt is also formed when a metal reacts with an acid.

**Semiconductor** A substance such as silicon that is neither a good conductor nor a good insulator. Transistors are made from semiconductors.

**Specific gravity** The density of a given volume of a substance, at a given temperature, compared with that of a similar volume of water at 39°F. Also called relative density.

**Spectroscope** An instrument for splitting up light into its various colors by means of a prism or a diffraction grating. The colors produced can then be studied.

**Surface tension** The property of the surface of a liquid that makes it behave as though it were covered with a thin elastic skin. It is caused by the forces of attraction between molecules in the surface of the liquid.

**Temperature** The degree of hotness of a substance which is measured on a temperature scale such as Fahrenheit or Centigrade.

**Thermostat** A device that regulates temperature. It cuts off the heating source if the temperature rises beyond a certain value, and vice versa.

**Transformer** A device for changing alternating current at one voltage to alternating current at a different voltage.

**Transistor** A semiconductor device that can amplify and control electric current. Transistors have taken the place of thermionic valves.

**Ultra-violet rays** Light waves that are of smaller wavelength than the visible light at the violet end of the spectrum. The Sun's radiation is rich in ultra-violet rays. They may also be produced artificially by special lamps.

**Vacuum** Empty spaces in which there are no atoms or molecules. A true vacuum cannot be obtained in practice, but the word is used to describe a space in which there is very little gas.

**Vapor** A gas that can be turned into a liquid by squeezing it without cooling.

**Wavelength** The distance from the crest of one wave to the crest of the next. Radio wavelengths are measured in feet.

**Weight** The downward force exerted by the gravitational pull on an object. The mass of an object does not vary, but the weight depends on how much gravitational pull there is. On Earth, for example, an object weighs more than it does on the Moon, where the gravitational pull is not so strong.

# *Index*